PRINTED CIRCUIT ENGINEERING

Optimizing for Manufacturability

PRINTED CIRCUIT ENGINEERING

Optimizing for Manufacturability

Raymond H. Clark

VNR VAN NOSTRAND REINHOLD
—————————— New York

Copyright © 1989 by Van Nostrand Reinhold
Library of Congress Catalog Card Number 88-5675

ISBN 0-442-21115-5

Printed in the United States of America

Van Nostrand Reinhold
115 Fifth Avenue
New York, New York 10003

Van Nostrand Reinhold International Company Limited
11 New Fetter Lane
London EC4P 4EE, England

Van Nostrand Reinhold
480 La Trobe Street
Melbourne, Victoria 3000, Australia

Macmillan of Canada
Division of Canada Publishing Corporation
164 Commander Boulevard
Agincourt, Ontario M1S 3C7, Canada

16 15 14 13 12 11 10 9 8 7 6 5 4 3 2 1

Library of Congress Cataloging-in-Publication Data

Clark, Raymond H.
    Planning the printed circuit manufacturing environment / Raymond
H. Clark.
        p. cm.
    Includes index.
    ISBN 0-442-21115-5 :
    1. Printed circuits—Design and construction. I. Title.
TK7868.P7C553 1988
621.381'74—dc19   88-5675

To all people involved in designing, engineering, manufacturing, inspecting, and procuring printed circuits.

Preface

I would like to present some definitions which will be helpful in understanding the purpose of this book. From *The American Heritage Dictionary of the English Language*:

Engineer
1. A person who skillfully or shrewdly manages an enterprise.
2. To plan, construct, and manage, as an engineer.
3. To plan, manage, and put through by skillful acts, or contrivance.

Engineering
1. The application of scientific principles to practical ends as the design, construction, and operation of efficient and economical structures, equipment and systems.
2. The profession of, or work performed by an engineer.

Some words encountered in the definitions of engineer and engineering are *Skillfully*, *Plan*, and *Manage*. This book is concerned with engineering the manufacture of printed circuit boards, and is dedicated to those people engaged in designing, planning, manufacturing, and achieving quality assurance in printed circuits.

In *The Handbook of Printed Circuit Manufacturing* (Van Nostrand Reinhold, 1985), I presented detailed practical and theoretical information on the operations involved in manufacturing printed circuits. It is possible to perform each operation in an optimum fashion, and still leave room for improvement. Much of that room for improvement requires the skillful application of scientific principles, planning, and management. It is the goal of this book to provide a sound background in industry standards and specifications, blueprint comprehension, artwork inspection, processes and tolerances, planning and quality assurance.

There is a wide gulf separating zero defect performance and 80% yield performance. The manufacturing (planning) engineer should be able to study a procurement specification, set of blueprints, and incoming artwork and determine how well a job will run in the shop. The same engineer must be able to formulate instructions for tooling (drill and router programming, photo, testing) and for manufacturing (drilling, plating, imaging, etching, etc.) that will enhance performance at each operation. Armed with an understanding of each operation's effect on the product and the tolerance which can be held, the engineer is in the best position to help the shop achieve the appropriate specification.

If the manufacturing or planning engineer lacks an understanding of what each operation can achieve, or lacks an understanding of industry specifications, artwork and blueprints, then the shop is in trouble. The final result is sensitively dependent upon the least skillful person handling the product. This book shows how to take measurements on artwork, and issue quantitative set-up instructions for tooling operations. Briefly, this book shows how to plan the manufacturing of the job through each operation.

During the last several years I have become increasingly aware of the need for a concise book on engineering/planning of printed circuits. It is my sincere wish that the people to whom this book is dedicated find it useful.

Raymond H. Clark
San Jose, CA

Contents

PRINTED CIRCUIT ENGINEERING

Optimizing for Manufacturability

Chapter 1

Introduction to Standards and Specifications

The printed circuit industry is fortunate to have wealth of information. Various military and government organizations and the Institute for Interconnecting and Packaging Electronic Circuits (IPC) publish an array of documents on all aspects of design, testing, end product material requirements, and processing for printed circuits. An individual need not be an expert to use this information. In fact, no one can be considered fully competent in his or her field of training and experience without a knowledge of the specifications and standards which have been established and are internationally recognized as industry guidelines.

The following are some of the reasons why it is necessary to have an understanding of industry standards:

1. To help us understand the difficulty and complexity of what we are designing and asking others to build.
2. To help us understand the difficulty and complexity of what we are being asked to build.
3. To help us understand whether a procurement or end product specification is reasonable.
4. To help us understand what is expected of us as designers, quality assurance personnel, planners, and manufacturers.
5. To help us develop a mature understanding of any aspect of our industry.
6. To allow us to obtain optimum results with the fewest number of failures.
7. To avoid overspecifying a product, test, or material requirement.
8. To ensure that what has been designed will meet form, fit, and function requirements.
9. To avoid shipping or accepting a product that is substandard.
10. To avoid rejecting material that is of good quality with respect to form, fit, and function.
11. To help determine if the artwork which has been rendered can be used to build circuitry that will meet electrical form, fit, and function requirements.

12. To provide reference information to help develop job quotations when reviewing blueprints, artwork, and procurement specifications.
13. To develop the necessary background for issuing instructions for all manufacturing operations.
14. To develop an understanding of what the industry is capable of reliably producing.
15. To obtain the knowledge for determining the absolute necessity for any given requirement.
16. To eliminate, as much as possible, conflicting and contradictory requirements.
17. To eliminate, as much as possible, unnecessary requirements.
18. To define terms, processes, and tests for common agreement among all personnel in the industry.
19. To help interpret design and manufacturing requirements.
20. To provide clear instructions on all processes, ranging from testing procedures to documenting and interpreting a blueprint.
21. To allow design, quality, and manufacturing personnel to benefit from the experience of people who have greater understanding of the subject matter than they do.
22. To provide a context for making wise decisions and issuing instructions.
23. To avoid having to issue repetitive instructions by relying on guidelines already established.
24. To reduce costs in design, manufacturing, and testing.
25. To be able to find accurate information and know that it has been used and proven reliable.
26. To help understand when information, advice, or instructions should be accepted or questioned.
27. To provide the reference materials for setting up design, quality, testing, and manufacturing training programs and department guidelines.
28. To avoid, as much as possible, making mistakes that waste the company's money.
29. To avoid, as much as possible, making mistakes that waste time and endanger the company's competitive position.
30. To avoid, as must as possible, making mistakes that demonstrate personal ignorance.

Tables 1-1 through 1-3 list many widely accepted government and nongovernment specifications and standards which are applicable to printed circuits. These documents are commonly cross-referenced among themselves, and in the specifications and standards of many companies. All organizations involved with printed circuitry should have copies of them.

It is not enough to rely solely on our own experience or the experience of those with whom we work. Printed circuits, electrical and electronic equipment,

Table 1-1 Government Specifications and Standards

FEDERAL

A-A-113	Tape, Pressure-Sensitive Film, Office Use
L-F-340	Film, Sensitized, Wash-Off Process, Diazotype, Moist and Dry Processes, Brownprint, Roll and Sheet
L-P-378	Plastic Sheet and Strip, Thin-Gauge Polyolefin
NN-P-71	Pallet, Material Handling, Wood, Stringer Construction, two-way and four-way Partial
QQ-C-576	Copper Flat Products with Slit, Slit and Edge-Rolled, Sheared, Sawed, or Machined Edges, (Plate, Bar, Sheet, and Strip)
QQ-N-290	Nickel Plating (electrodeposited)
QQ-S-571	Solder, Tin Alloy, Tin-Lead Alloy, and Lead Alloy
QQ-S-781	Strapping, Steel, and Seals
QQ-S-566	Box, Folding, Paperboard
PPP-B-601	Boxes, Wood, Cleated Plywood
PPP-B-621	Box, Wood, Nailed and Lock Corner
PPP-B-636	Box, Shipping, Fiberboard
PPP-B-676	Boxes, Setup

MILITARY

MIL-D-8510	Drawings, Undimensioned, Reproducibles, Photographic and Contact, Preparation of
MIL-C-14550	Copper Plating (Electrodeposited)
MIL-P-116	Preservation, Methods Of
MIL-P-13949	Plastic Sheet, Laminated, Metal Clad (For Printed-Wiring Boards), General Specification for
MIL-F-14256	Flux, Soldering, Liquid (Rosin Base)
MIL-G-45204	Gold Plating (Electrodeposited)
MIL-I-45208	Inspection System Requirements
MIL-I-46058	Insulating Compound, Electrical (For Coating Printed Circuit Assemblies)
MIL-P-55110	Printed-Wiring Boards
MIL-P-81728	Plating, Tin-Lead (Electrodeposited)
MIL-Q-9858	Quality Program Requirements
MIL-STD-105	Sampling Procedures and Tables for Inspection by Attributes
MIL-STD-129	Marking for Shipment and Storage
MIL-STD-130	Identification Marking of U.S. Military Property
MIL-STD-147	Palletized Unit Loads
MIL-STD-202	Test Methods for Electronic and Electrical Component Parts
MIL-STD-275	Printed Wiring for Electronic Equipment
MIL-STD-794	Parts and Equipment, Procedures for Packaging and Packing of
MIL-STD-810	Environmental Test Methods and Engineering Guidelines
DOD-100	Engineering Drawing Practices
DOD-STD-1000	Drawings, Engineering and Associated Lists
DOD-STD-1686	Electrostatic Discharge Control Program for Protection of Electrical and Electronic Parts, Assemblies, and Equipment
MIL-STD-45662	Calibration Systems Requirements

Table 1-2   IPC S-100 Standards Manual: Table of Contents

Table 1-2 *(Continued)*

Joining Methods and Techniques

IPC-S-804	Solderability test methods for printed wiring boards (supersedes IPC-S-801 and IPC-S-803)
IPC-S-805	Solderability test for component leads and terminations
IPC-S-815A	General requirements for soldering of electronic interconnections

Rigid 1 and 2 Sided

IPC-D-319	Design standard for rigid single and double sided printed boards C.B.D.
IPC-D-320A	End product specification for single and double sided printed boards
IPC-TC-510A	Procedure for installing and inspecting clinched wire type interfacial connections in rigid printed circuit boards.
IPC-TC-550A	Procedure for the design, assembly, and testing of fused-in-place interfacial connections in rigid printed circuit boards

Rigid Multilayer

IPC-ML-910A	Design standard for rigid multilayer printed boards
IPC-ML-950B	Performance specifications for multilayer printed wiring boards
IPC-ML-975	End product documentation specification for multilayer printed wiring boards

Flexible 1 and 2 Sided

IPC-FC-240C	Specification for single-sided flexible wiring boards
IPC-D-249	Design standard for single and double sided flexible boards C.B.D.
IPC-FC-250A	Specification for double sided flexible wiring with interconnections

Flex and Rigid-Flex Multilayers

IPC-RF-245	Specification for rigid/flexible printed boards C.B.D.
IPC-ML-990	Performance specification for flexible multilayer wiring

Connector

IPC-FC-217	General document for connectors, electric, header/receptacle, insulation displacement for use with round flat cable
IPC-FC-218B	General specification for connectors, electric flat cable type
IPC-FC-219	Environmentally sealed flat cable connectors
IPC-C-405	General document for connectors, electric, printed wiring board

Flat Cable

IPC-FC-202	Specification for flat cable flat conductor shielded C.B.D.
IPC-FC-203	Specification for flat cable, round conductor shielded C.B.D.
IPC-FC-210	Performance specification for flat-conductor cable, (Type FCC) undercarpet, power C.B.D.
IPC-FC-211	Performance specification for undercarpet flat cable, data transmission C.B.D.
IPC-FC-213	Performance specification for flat undercarpet telephone cable C.B.D.
IPC-FC-220C	Specification for flat cable, flat conductor, unshielded
IPC-FC-221A	Specification for flat-copper conductors for flat cables
IPC-FC-222	Specification for flat cable, round conductor, unshielded

Table 1-2 (*Continued*)

Alternate Interconnections		
IPC-BP-421	General specification for rigid printed board back-planes with press-fit contacts	
IPC-DW-425	Design and end product requirements for discrete wiring boards	
IPC-DW-426	Acceptability requirements for discrete wiring assembles	C.B.D.
Other specifications		
IPC-DR-570	Specification for $\frac{1}{8}''$ diameter shank for printed boards	
IPC-HM-860	Qualification and performance of multilayer hybrid circuits	C.B.D.

C.B.D.-Currently being developed. The technical committees of the IPC are presently developing this standard. After this standard is approved, it will automatically be distributed to all manual holders with updating services.

Table 1-3 Other NonGovernment Documents

American National Standards Institute (ANSI)	
ANSI-Y14.5M	Dimensioning and Tolerancing
ANSI-Y14.1	Drawing Sheet Size And Format
ANSI-Y32.16	Reference Designations for Electrical and Electronic Parts and Equipment
American Society For Testing And Materials (ASTM)	
ASTM-E53	Standard Method for Chemical Analysis of Copper (Electrolytic Determination)
ASTM-B-487	Measuring Metal and Oxide Coating Thickness by Microscopic Examination of a Cross Section

chemicals, materials, and processes are being developed, used, and manufactured in many cities, states, and nations around the world. We cannot limit our understanding of printed circuits to our own tiny realm. To do so is to do ourselves and our employers a disservice. By studying industry specifications and standards, we can make better products, less expensively, and more competitively than before.

Designers, quality assurance people, planners, and manufacturing personnel must know what the industry considers standard, doable, and necessary. The documents listed in Tables 1-1 to 1-3 contain the information we must know and be able to use. Our jobs and the economic viability of our company depend on it. Unnecessary requirements add cost and needless delays. Necessary requirements which go unforseen waste time and money. There are times when one must push the limits of what can be reliably accomplished. When there is such a need, designers, quality assurance people, planners, and manufacturing personnel should press for it. By contrast, there are times when a requirement is unnecessarily stringent and serves no purpose. When nothing is to be gained by such a requirement, design, quality assurance, planning, and manufacturing people should come to an agreement and eliminate it.

Without a knowledge of industry specifications and standards, the opportunity to include unnecessary requirements or to miss necessary and forseeable ones is always present. The specifications and standards listed in the Tables 1-1 to 1-3 are often referenced on blueprints, purchase orders, and company specifications. It is generally assumed by people who use these documents that the reader understands or has access to them and will refer to them as needed.

CORPORATE PROCUREMENT SPECIFICATIONS

Corporations often write their own printed circuit procurement specifications. They do this for many of the same reasons that all standards and specifications are written. Procurement specifications are an important part of the quality plan. These specifications inform vendors and inhouse personnel of what is required, whether the corporation is a single plant or a multiplant, multi-billion-dollar giant.

Printed circuit manufacturers cannot afford to disregard the specifications of their customers. Specifications are considered a part of all blueprints issued by the corporation. It is the job of the planner to read all procurement specifications and transmit them to the rest of the shop in order to assure compliance. Planners must also know when any of the requirements in a specification do not make sense or are unusually restrictive. For instance, if a procurement specification (or blueprint) has a requirement for plated-through hole diameters with a tolerance of $\pm.001$ inch, the planner must raise a warning flag. Such a requirement should be reviewed with the customer. It is better to question unreasonable requirements prior to commencing work. Few tasks are more displeasing and make for poorer customer–vendor relations than accepting unreasonable requirements, failing to meet them, and then asking the customer for a waiver of the out-of-spec condition. The planner is in the good position of being able to protect the best interests of his/her own company and the interests of the customer at the same time.

Most procurement specifications contain the following information:

1. General
   a. The purpose of the specifications and the items covered. Many corporations have multiple procurements specifications, each covering a different type of printed circuit: double-sided, multilayers, controlled impedance multilayers, fine-line or three-track circuits, etc.
   b. The order of precedence. The blueprint takes precedence over the specification, and the purchase order takes precedence over everything.
   c. Underwriters Laboratories (UL) requirements.

  d. Military requirements.

  e. Reference documents. These include military, IPC, and other industry specifications and standards which are referenced and form a part of the procurement specification.

 2. Master Pattern (Artwork)

  a. The customer will furnish a 1:1 scale master pattern of all layers.

  b. The customer is responsible for assuring the accuracy of the master pattern. However, the vendor is responsible for assuring that the artwork is suitable for producing printed circuits in accordance with the blueprint, the specification, and the purchase order.

  NOTE: These are important words. Planning and quality assurance personnel must be aware of this common requirement.

  c. Discrepancies are to be reported to the customer.

  d. No changes may be made to the master pattern without written approval by the customer.

 3. Laminate and Multilayer Construction

 Here the type of laminate (such as FR-4), the thickness, and copper cladding are spelled out. Sometimes there are notes on minimum dielectric requirements, layer sequencing, and registration for multilayer printed circuits.

 4. Plating

 This section includes the types and thicknesses of all plated metals. Notes indicating that the board is to be made by the solder mask over bare copper procedure are often found here.

 5. Conductors

  a. The amount of growth or reduction of conductive patterns.

  b. Minimum trace width and air gap.

  c. Types and degrees of allowable defects for inner and outer layer conductors.

  d. Minimum annular rings for inner and outer layers.

  e. Appearance of plated metals and reflowed tin-lead.

  f. Types of allowable repairs.

 6. Plated holes

  a. Thickness of plated metals.

  b. Requirements for smear removal or etchback.

  c. Type and degree of allowable defects.

7. Dimensions
   a. Finished board thickness.
   b. Allowable warp and twist.
   c. Bevel and chamfer of corners and contact fingers.
   d. Radii of slots, corners, and internal cutouts.

8. Solder mask and Legend
   a. Type and color of solder mask allowed.
   b. Type and degree of allowable defects.
   c. Adhesion and cure testing requirements.
   d. Type and color of epoxy legend ink allowed.
   e. Requirements for legibility and accuracy of registration.
   f. Side(s) of the board on which solder mask and legend ink may be applied.

9. Markings
   a. UL marking requirement: Type designation and 94V-0 or 94V-1.
   b. Date code type (four-digit week/year, month/day/year, etc.)
   c. Serialization requirements and military markings such as those of the Federal Supply Code for Manufacturers (FSCM).
   d. Revision level.
   e. Location of indicated markings.

10. Testing and Inspection
    a. Bare board continuity.
    b. High-voltage breakdown.
    c. Solderability.
    d. Microsectioning.
    e. Sampling plan to be followed by the vendor.
    f. Documentation and records.
    g. Certification of compliance with specification.

11. Packaging
    a. How boards are to be bagged: how many in a bag, what type of bag, how bags are to be sealed, etc.
    b. Boxing and packaging materials.
    c. Labeling requirements.

This list is typical of the information found in procurement specifications. Some corporations go into greater or less depth than others. Obviously, the planner must know the commonly referenced industry standards in order to be

CUSTOMER _____   SPECIFICATION _____

DATE CODE: SOLDER MASK:
LOGO: LEGEND:
U.L.   COPPER:
LAMINATE – D/S: NICKEL:
        – M/L: GOLD:
                                      SOLDER:
MIN. DIELECTRIC: SMOBC REQUIRED:
ELECTRIC TEST: MIN. TRACE WIDTH:
OVERAGE ACCEPTED: MIN. AIR GAP:
                                      MIN. ANN.RING – INNER:
                                                    – OUTER:

OTHER INFORMATION:

_____
_____
_____
_____
_____
_____
_____
_____
_____
_____
_____
_____
_____

PACKAGING AND SHIPPING REQUIREMENTS:

_____
_____
_____
_____
_____
_____
_____
_____

Fig. 1-1. Example of a specification summary sheet. The most important information from a corporation specification can be summarized on one page. When a corporation has no specification, information can be filled in as more is learned about that company. The specification summary sheet thus can be used as an informal summary of the likes, dislikes, and requirements of the customer.

able to evaluate effectively any of the requirements listed in a corporate specification.

In addition to knowing the requirements expressed in the corporate specification, the planner must disseminate this information to other departments at the manufacturing plant. These other departments usually deal with final and presolder mask inspection and artwork inspection. The shipping department must also be aware of the packaging requirements. An easy way to disseminate the information in a procurement specification is to fill out a one-page summary or condensation (see Figure 1-1). Copies of the specification summaries can be kept in binders in each of the relevant departments. A copy can even be placed in the traveler package to accompany the job throughout the manufacturing process. If there is a serious problem on a job, the full specification may be consulted.

Appendixes A–F discuss common industry standards which planners must know.

Chapter 2

Understanding Blueprints

The purpose of a blueprint is to provide the information necessary to build a printed circuit board. Printed circuit boards are complicated products. Multilayer printed circuits are often the most difficult to design, build, and test of all components in a piece of electronic equipment. For this reason, all persons involved with the printed circuit process must know what information is required for a correctly documented printed circuit blueprint. Incorrectly and incompletely documented blueprints waste time and money, and may have a severe negative impact on the timely marketability of a product and the economic viability of a company. Blueprints and the information they contain are therefore not to be taken lightly.

For every item required but not included on the blueprint, there is the possibility that the printed circuit may not fully meet the expectations of the end user. For any given printed circuit, certain information is absolutely required, such as hole sizes and fabrication (board outline) dimensions. There is also other information which, if not documented on the blueprint, may result in the product not coming out the way it is desired or needed (board thickness, laminate type, solder mask, metal thickness, etc.). Information required but not documented on the blueprint must be decided upon by the planner at the printed circuit manufacturing facility. The planner can use three methods in deciding how to proceed when writing the manufacturing instructions on the job traveler. Design and quality assurance people must understand that information they fail to include on a blueprint will come from one of the following three sources: (1) the customer (end user or designer), (2) the customer's end product/procurement specification, or (3) guesswork.

Each of these methods of gathering required information has associated drawbacks. It should be obvious that taking a guess or making an assumption is the least reliable method. It is common for planners to assume certain information when it is not on the blueprint, such as FR-4 laminate. .062-inch board thinkness, and .001-inch of copper plating followed by tin-lead. Most assumptions are based on meeting the most frequently required specification. However, these assumptions may not result in a usable printed circuit board. Some assumptions can be even more risky, such as layer sequencing, copper cladding of internal layers, designating holes as plated or nonplated, and assigning hole size tolerances. In making assumptions, the only thing that is perfectly clear is that neither the manufacturer nor the designer can be absolutely certain that the resulting printed circuit will be usable.

Another source of information to complement the blueprint and the manufacturing instructions is the end product or procurement specification of the company buying the printed circuit board. As discussed in Chapter 1, companies write specifications so that there is no misunderstanding on what is required in most circumstances. The purchase order, and then the blueprint, take precedence over the company's specification. However, when needed information is missing from the blueprint, it is permissible to consult the specification as a source document. The specification actually becomes part of the blueprint and should be treated accordingly. Thus, no matter how much information is presented on the blueprint, the company's procurement specification must be referred to routinely when planning, inspecting, testing, or designing a printed circuit board. This should be done even before reviewing the requirements presented on the blueprint. Planners and quality assurance people should understand that when a company has a procurement specification, the board manufacturer is expected to meet the requirements of the specification and the blueprint. Consulting the specification, if there is one, will eliminate unnecessary telephone calls that do little more than demonstrate the vendor's ignorance to the customer. However, there is a drawback in relying too heavily on the information in a specification. Sometimes this information may conflict with other standard practices of the company. Not all printed circuit designers know the procurement specifications of their own company. Furthermore, the standard information (dielectric spacing, minimum trace width, layer sequencing, hole tolerances, air gaps, etc.) may not be adequate for the printed circuit to meet the functional requirements for which it is designed.

There are times when calling the customer is unavoidable and therefore justified. It is the job of planners and quality assurance personnel to know the requirements. It is not their job to make assumptions when the facts are not present. Generally, when the printed circuit manufacturing planner has to call the customer, the information received is reliable. However, placing the planner in the situation of having to ferret out missing information creates delays in releasing the job to the manufacturing department. There is also the likelihood of changes being made to the blueprint or artwork which are not fully documented. This can place the customer and the printed circuit manufacturer at odds when the delivered product does not fully meet the customer's expectations.

When information is missing from the blueprint, uncertainties exist. Since the missing information must come from one of the three sources discussed above, there is a potential for discrepancies and delays, which result in poor customer–vendor relations and inefficient operations on the part of both parties.

Thus, the best solution is to provide clear and complete blueprints. Designers, buyers, and manufacturers of printed circuits must not underestimate the potential for problems when blueprints are not properly documented.

INFORMATION INCLUDED ON THE BLUEPRINT

1. Title Block. The title block is usually located at the bottom right-hand side of the blueprint (see Figure 2-1). While companies differ in the amount of information they list there, the following information is most commonly found:
   a. Company name/logo
   b. Title of the part shown
   c. Part number
   d. Revision level
   e. Tolerances: Fabrication (board outline)
      (1) Hole sizes
      (2) Angular
   f. Laminate type
   g. Signatures of approval

2. Part Number. The part number often consists of several sets of numerals, such as 45-122345-001. The person reading the blueprint should always refer to the entire number. Many people drop sets such as 45- or -001. It should always be remembered that the more attention is paid to details, the less likely it is that the wrong part number will be built. Sometimes there is no number, only a name for a part. The blueprint may consist of several drawings, together with a separate parts list and a directory of hole sizes and locations. All of these items make up the blueprint. Therefore, all of these items must have the identical part number and revision level. There are instances in which these numbers are not identical. In such cases, there must be a note on the blueprint which informs the reader of this fact. The note must also indicate the part or revision numbers. The individual reading the blueprint must verify that all part and revision levels are correct.

3. Revision Level. The need for accuracy regarding the part number also applies to the revision level. The revision levels on the various documents of the blueprint must agree. No documents with unmatching revision levels should be considered reliable. Documents with differing revision or part numbers must be questioned; no work should proceed until this basic discrepancy has been resolved.

UNLESS OTHERWISE SPECIFIED	APPROVALS	SIGNATURE AND DATE		
DIMENSIONS ARE IN INCHES	DRAWN			
TOLERANCES				
FRACTIONS	DECIMALS	ANGLES	APPROVED	
±1/64	XX ±.02	±1°	APPROVED	
	XXX ±.010		APPROVED	
125/				
ALL SURFACES √ EXCEPT AS NOTED	APPROVED			
PROJECTION	APPROVED			
NOTE: DEBURR AND BREAK ALL SHARP EDGES EXCEPT AS NOTED.	CONTR			
DO NOT SCALE THIS DRAWING				

MASTER DRAWING -
SURFACE MOUNT
TEST BOARD

SIZE	FSCM NO.	DWG NO.		REV
D	15280	59-08		—
SCALE 2/1	RELEASE DATE		SHEET 1 of 5	

NONE	±R&D	
NEXT ASSEMBLY	USED ON	
APPLICATION		
WORKMANSHIP PER MIL–STD–454 REQUIREMENT 9		

Fig. 2-1 The title block.

15

4. Signatures of Approval. Most blueprints contain an area near the title block for signatures of approval. The signatures must also be dated. Generally, blueprints are considered incomplete and/or inaccurate when they lack approvals.

5. Material Type, Thickness, and Cladding*. The blueprint must identify the type of laminate upon which the printed circuit is to be built. There are major cost and functional differences between the available laminate types. Blueprints which do not call out the laminate type are typically run on FR-4/GF material. Common types of laminate are as follows:

   a. GF or FR-4: This is general-purpose, flame-retardant epoxy/fiberglass material. It is the most commonly used laminate in the United States. The designations GF and FR-4 have the same meaning. GF is the military designation, and FR-4 is the National Electronics Manufacturing Association (NEMA) designation. It typically carries a UL flammability rating of 94V-O, which is the best UL flame-retardant rating.

   b. GE or G10: This is general-purpose epoxy/fiberglass material; it is not flame retardant. GF material has almost completely replaced GE material. Any blueprint calling for GE should almost automatically be considered to call for GF, unless there is a specific note not to make this substitution. GE material carries a poor UL flame-retardant flammability rating.

   c. GH: This is epoxy/fiberglass laminate used in high-temperature applications where mechanical strength is important.

   d. GI or polyimide: This is a polyide/fiberglass laminate. It is used in high-temperature applications, especially for military applications. It is one of the most expensive laminates, but is much less expensive than Teflon materials.

   e. GX, GT, and GR: This is Teflon/fiberglass laminate used for microwave applications. It is one of the most expensive laminates available. It is soft and is more difficult to work with than other laminates.

   f. CEM-1 and CEM-3: These are epoxy/fiberglass laminates over paper core materials. They are less expensive than most other materials, including GF.

   g. FR-2: This is fire-retardant phenolic resin laminate over paper. It is an inexpensive material.

   Thickness and foil cladding are fully discussed in the Appendix E. However, on blueprints, outer layer foil is sometimes called out as a finished thickness of 2-oz or 3-oz copper. This means starting with 1-oz foil and

---

*For a detailed discussion of laminates, see the *Handbook of Printed Circuit Manufacturing*, Section 1.

plating to 2 oz or starting with 2-oz foil and plating to 3 oz, since 1 oz of copper (.0014 inch) will typically be plated for .001 to .0015 inch of hole wall copper plating.

Thickness may be shown as a dimension on a drawing or as a number (.062 inch) in the material callout section of the title block. The thickness requirement of a single-sided or double-sided board is simply the laminate callout. The thickness requirement of a multilayer printed circuit is a little more difficult. If the printed circuit board has contact fingers, the thickness is generally understood to mean at the contact fingers, since these must fit into a connector of some kind. If there are no contact fingers, the thickness will typically be measured metal-to-metal at the outer layers after plating. The planner must keep this in mind because he/she must plan the construction of the printed circuit. This will be fully discussed in Chapter 5.

6. Number of Layers and Layer Sequence. Blueprints must clearly show the number of layers for which the printed circuit is designed. Sometimes this information is left off. When this is the case, there are clues we can look for. If the laminate callout is for .062 1/0, we know that we are dealing with a single-sided printed circuit. If the callout is for .062 1/1, we are probably dealing with a double-sided printed circuit. However, we may also be dealing with a multilayer printed circuit. A common blueprint error is to call out a thickness as .062 1/1, when what is really meant is simply .062.

If we are looking at a blueprint which clearly indicates a multilayer, we must be able to determine the layer sequence. A diagram showing layers of circuitry must indicate the layer numbering or name of each layer (see

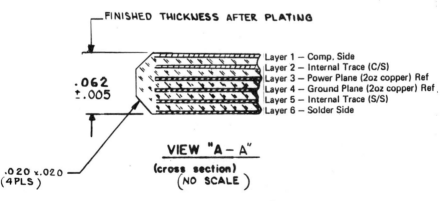

Fig. 2-2 A layer sequence diagram.

Figure 2-2). In addition, the side showing on the blueprint must identify which side of the board (which outer layer) is being viewed. If the side showing is not identified, there is a possibility that the layer sequencing on the final product may come out backward, for example, 6, 5, 4, 3, 2, 1 instead of 1, 2, 3, 4, 5, 6. If the side is identified as "Component," the component side must be identified as layer 1 or layer 6 for a six-layer printed circuit.

7. Dimensions and Tolerances. The dimensions of the printed circuit are one of the most important items on the blueprint. The designer and planner must verify that all dimensional requirements have been accounted for. Dimensions which must be present include the following:
   a. Overall board outline.
   b. Slot location, depth, and width.
   c. Corner chamfers.
   d. Contact fingers: Length and width of the area to be gold plated. This includes the tape line for gold plating.
      • Bevel—angle and recess.
   e. Internal cutouts, dimensions, and locations.
   f. Dimensioned holes for tooling or mounting requirements.
   g. Board thickness.
   h. Dielectric spacing between layers of circuitry for multilayer printed circuits.
   i. Areas of relief in solder mask.
   j. Location and spacing of test coupons.

   It is common for blueprints to have one or more of these items missing. The planner, designer, quality assurance person, or manufacturing person using the blueprint must verify that all are present. Not only must the dimensions be present, they must be correct and have manufacturable tolerances. Thus, the planner must also pay attention to tolerances. Planners should be familiar with the tables presented in MIL-STD-275 and IPC-D-300. It is also common for dimensions to be in error. Some common errors are:
   1. Slots, cutouts, and large holes breaking into circuitry.
   2. Individual dimensions adding up to more than the overall dimension.
   3. Dimensioned tooling/mounting holes not referenced from a datum which is common to board outline features.
   4. Tolerances on dimensions unrealistically tight (e.g., board outline tolerance tighter than ± .005 inch).

   Ideally, dimensions are listed on one page of the blueprint. Sometimes, however, they are spread over two or more sheets of a blueprint set. Planners must be aware of the potential for problems, as when board outline dimensions are listed on one page and nonplated tooling/mounting holes on another. It is good planning practice to note on the page with board outline dimensions that dimensioned holes are listed on a second page. This will prevent a less observant drill programmer from using the artwork to locate these nonplated holes incorrectly.

Fabrication tolerances may be printed next to the dimensions on the blueprint or may be listed in the title block. The planner must check for dimensions listed in both locations. Title blocks typically list dimensional tolerances as:

.x = ±.020 inch.
.xx = ± .010 inch.
.xxx = ± .005 inch.

8. Hole Sizes and Tolerances. Hole size requirements are generally listed in a hole chart (see Figure 2-3). The hole chart must list the following information about each hole:

a. The code used to designate each hole on the blueprint.

b. The nominal size and tolerance for each hole in the chart. Sometimes a range is shown instead.
    EXAMPLE: A B-coded hole may be listed as .075 ± .005 inch or as 070 to 080 inch.

c. Whether the hole is to be plated or nonplated, or whether plating is optional.
    Planners should be aware of holes which are not shown on the hole chart. Sometimes holes are shown only on the blueprint and are dimensioned in. Holes like these should be redlined into the hole chart by the planner.

9. Details. Many blueprints contain items called "details." Details are iso-

HOLE SCHEDULE				
PAD DIA.	CODE	DESCRIPTION	QTY	PLATED
.050	A	.028 DIA. ± .003	534	YES
.060	UNMARKED	.039 DIA. ± .003	1813	YES
	B	.059 DIA. ± .003	1	NO
	C	.125 DIA ± .002	3	NO
	D	.136 DIA ± .003	2	NO
	E	.156 DIA ± .003	1	NO

Fig. 2-3   A hole chart.

lated sections of a blueprint which have been blown up in scale for clarity (see Figure 2-4). Common uses of details include:

a. Beveling and chamfering requirements.
b. Dimensioning of slots at contact fingers and internal cutouts.
c. Positioning and dimensioning of holes in a complicated pattern. Some hole patterns are shown as a mark on the blueprint and the mark is labeled as "Detail A," "Detail B," etc. The detail of A or B will then show all features of the holes. Usually one hole in the detail will serve as a secondary datum from which the other holes are dimensioned.
d. Locations and dimensions of pads in a repetitive surface mount pad group.
e. Graphic display of multilayer construction (copper cladding and dielectric spacing) requirements.
f. Orientation of hardware or eyelets which are to be installed.
g. Location of edge plating or of selective plating of any kind.
h. Countersinking requirements for holes.
i. Isolated areas where fabrication or drilling tolerances are especially critical.
j. Information on test coupons.
k. Any type of information which would clutter up the blueprint if presented with the other information.

10. Notes on the Blueprint. Notes are a very important part of blueprints. They give all the information needed to complete the specification of the part. There is almost no end to the amount of information which can be presented as a note on a blueprint. A note may be as simple as one referring the reader to a specification to be followed for all fabrication requirements. In this case, the specification becomes an integral part of the blueprint. No one can plan or build a printed circuit correctly without studying all of the notes. Items which are commonly noted on blueprints include:

a. Types and thickness of metals to be plated.
b. Special process requirements, such as solder mask over bare copper (SMOBC), or all-gold body, or selective plating of any metal.
c. Minimum trace width and air gap, or minimum annular ring for inner and outer layers.
d. Etchback requirements for multilayers.
e. Type of laminate and thickness of board, as well as minimum dielectric spacing requirements between layers.
f. Where board thickness is to be measured (i.e., at the contact fingers or across (metal surfaces).
g. Type of solder mask, color, and number of sides. The same information can be given for epoxy legend nomenclature as well.
h. Grid requirements for hole drilling.

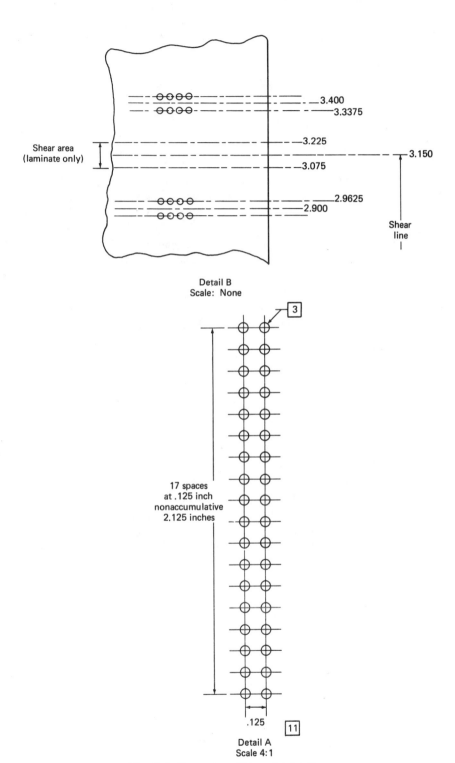

Fig. 2-4 Examples of blueprint details.

    i. Hardware requirements.

    j. Testing requirements: electrical, environmental, Group A or Group B military requirements.

    k. Which side of the printed circuit is shown in the drawing.

    l. Marking requirements, such as date code, serialization, and UL requirements.

    m. Sometimes the part number and revision levels between the part number, blueprint, and artwork do not agree. When this happens, a note is made on the blueprint specifying which part and revision level to use.

    n. Refer to other sheets in a blueprint set for dimensioned tooling holes or further fabrication requirements.

    o. Restrictions on solder mask or epoxy legend ink. Notes such as those dealing with the solder mask to be relieved in indicated areas or no solder mask on pads are common.

    p. Other specifications to be followed unless otherwise noted.

    Sometimes important information is communicated by vague notes. The type of laminate, for instance, may be called out by referring to an internal or industry specification. The planner must be familiar enough with industry specifications, or with the specification of the customer, to realize when important information is being transmitted.

11. Other sheets in a multisheet blueprint set may depict the circuitry on each layer. This is a requirement of MIL-STD-275.

    A blueprint contains much more information than dimensions and hole sizes. The planner, designer, and quality assurance person should study every square inch of a blueprint for information that may not be obvious. When information seems vague or is presented by reference to another document, a warning flag should go up in the mind of the reader. Vagueness and references to other documents cannot be passed over by anyone involved in planning, building, or inspecting the printed circuit board.

Chapter 3

Processes and Tolerances

In Appendixes A, B, C, D, E and F, the reader is introduced to several of the cornerstone documents of the printed circuit industry. These documents guide much of the design, testing, and acceptability requirements which are presented in artwork, blueprints, and corporate procurement specifications. In other words, these documents have a major impact on the work of the printed circuit manufacturer. The manufacturer must be familiar with them in order to know what is expected. This material also gives the reader a firm background from which to evaluate the requirements of any given job, as delineated by the artwork, blueprint, procurement specification, and purchase order.

Anyone who inspects artwork, or who reads a blueprint or specification, must be able to distinguish between the following:

1. Requirements which are recognized and accepted by the electronics industry and requirements which exceed or conflict with recognized and accepted standards.
2. Requirements which meet various industry standards but which will cause other problems downstream.

Without a good knowledge of industry wide recognized standards and specifications, the reader will find it difficult to make sense of a set of requirements. It is one thing to commit the resources and reputation of the company to a requirement with a full understanding of what is in store, but quite another when no such understanding exists.

With the information on specifications presented in Chapter 1 and Appendix A, B, C, D, E and F, the reader possesses the background to evaluate the requirements of any given printed circuit. The purpose of this chapter is to help the reader understand the manufacturing operations and the processing windows through which the printed circuit must pass as it is being produced. There are many actions which the planner and other shop personnel can take to help build a product which meets all specifications in the fullest sense of the term: conformance to requirements.

Many customers expect the printed circuit planner to be able to spot flaws in their designs. In other words, planners are expected to have a knowledge of industry standards and a good deal of common sense about printed circuit manufacturing. The planner must understand this and must realize that much of the artwork and blueprints provided to his/her company will have to be carefully reviewed. Industry standards have evolved from an understanding of processing tolerances and capabilities and the need to obtain certain end results. This chapter deals with processing tolerances, processing capabilities, and desired end results. There is a good deal of repetition of information. This is done on purpose. No printed circuit manufacturing operation is an island unto itself. Almost every manufacturing decision will have an impact on one or more of the processes downstream. Below is an outline showing how one operation can be affected by another.

1. Areas affected by artwork
   a. Photo
   b. Programming (and all aspects of N/C processing)
   c. Inner layer imaging
   d. Outer layer imaging
   e. Solder mask and legend
   f. Fabrication
      • Contact finger center line
      • Centering of contact fingers
   g. Electrical testing

2. Areas affected by drilling
   a. Drilling and fabrication
   b. Imaging: inner and outer layers, all facets
   c. Solder mask and legend
   d. Plating—electroless copper
      (1) Smear removal for double-sided panels
      (2) Electrolytic copper
      (3) Contact fingers
   e. Group A inspection and microsectioning
   f. Electrical testing
      (1) Hole location
      (2) Drilling into ground planes

3. Areas affected by imaging choices
   a. Plating
      (1) Circuitry
      (2) Effects of plating baths

  b. Etching
   (1) Trace width
   (2) Poor-resist strip/overetching
  c. Resist stripping—type of resist
  d. Pollution control system
  e. Electrical testing and inspection
  f. Drilling—hole tenting, desired or not

4. Areas affected by plating
  a. Solder mask and legend
   (1) Plating height
   (2) Metal overhang
  b. Fabrication—second drill and routing, relief requirements
  c. Photo
  d. Imaging
   (1) Effects on circuitry
   (2) Effects of unfriendly baths (Sn-Ni, gold, pyro copper)
  e. Pollution control system
  f. Electrical testing
  g. Group A and inspection requirements
  h. Multilayer lamination—thickness requirements

5. Areas affected by etching
  a. Photo
  b. Imaging
  c. Plating—etch resist
  d. Pollution control system
  e. Electrical testing
  f. Group A requirements

6. Areas affected by solder mask and legend application
  a. Scrubbing and cleaning
  b. Plating—height, hole plugging
  c. Photo
  d. Programming—panel size and layout
  e. Surface mounting requirements
  f. Electrical testing
  g. Group A requirements
  h. Fabrication—reliefs needed for routing

7. Areas affected by electrical testing
  a. Photo
  b. Imaging

    c. Plating
      (1) Overplating
      (2) Overhang
    d. Imaging choice
    e. Fabrication
    f. Programming and drilling

Many of the people and organizations designing printed circuits today have no acquaintance with IPC and military specifications. One of the purposes of this book is to help planning, design and quality assurance personnel develop a working, knowledge of these industry standards. Nothing is learned if standards and specifications are ignored when studying process operating windows. In addition, little is gained in studying process operating windows without an understanding of what requirements the printed circuit manufacturer is expected to attain according to recognized industry standards.

Throughout the discussion on process operating windows, references will be made to IPC and/or military standards and specifications. Many blueprints and corporate procurement specifications have requirements which range from contradictory and difficult to impossible. Planners, designers, and quality assurance people who have a working knowledge, of accepted industry standards, and of processing windows, are in a position to evaluate properly and understand any set of requirements. These people occasionally encounter requirements which are out of step with IPC and military requirements. They perform a real service to themselves and to their customer or vendor when they have the knowledge to question these requirements.

DRILL BIT SELECTION FOR PLATED-THROUGH HOLES

Many factors affect the finished hole size and the proper drill bit selection to achieve the desired hole size. The single most important consideration is the desired finished hole size and the tolerance allowed. Because it is important to understand the factors which affect holes sizes and tolerances, this item will be discussed last. The reader should, however, keep in mind the hole sizes and tolerances listed in Appendix tables B-7 (IPC-D-300G) and A-2 (MIL-STD-275E) while reading the material in this section. Tolerances are listed in MIL-STD-275E as: $\pm$.003 inch is a spread of .006 inch. Notice that there are three classes: Preferred, Standard, and Reduced Producibility. The tightest tolerance which the board manufacturer is called upon to hold in IPC and military standards is $\pm$.002 inch (spread of .004 inch). Even this is considered Reduced Producibility. Tight tolerances have a negative impact on yields, costs, and profits, as well as on the ability to procure or deliver printed circuits at low cost with on-time delivery.

1. Type and Thickness of Metals to be Plated. Drill bit selection must reflect the amount of metal to be plated. The basic rule of hole size is that each .001 inch of plated metal will reduce the hole .002 inch. (See Figure 3-1 for an explanation of how plated metal affects hole size.) The planner must know the customer's requirement. Most plated holes require .001 inch of nominal thickness of copper. A requirement for .0015 inch is probably the next most common requirement for copper thickness. There are blueprints which call for much more than .0015 inch. These requirements must be thought out when choosing the drill size.

   Tin-lead and other metals, such as nickel, tin, and tin-nickel, also contribute to a reduction in hole size when they are pattern plated over the copper. Refer to Table 3-1 for suggested drill sizes as a function of copper, solder, and hole tolerance. The nominal hole size is the value in the center of the tolerance spread.

2. Panel Thickness and Aspect Ratio. Aspect ratio is defined as the ratio of board thickness divided by the diameter of the smallest plated-through hole.

$$\text{Aspect ratio} = \frac{\text{board thickness}}{\text{diameter of smallest hole}}$$

For panels which are .062 inch or thinner, the aspect ratio is of little concern. Aspect ratio becomes of greater concern as panel thickness increases

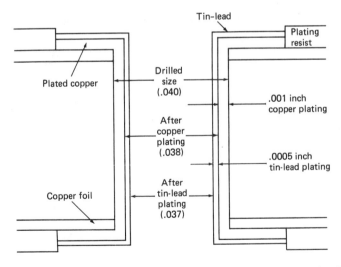

Fig. 3-1   The basic rule of hole size is that each .001 inch of plated metal reduces the hole diameter by .002 inch.

Table 3-1 Drilling Allowances for Plating
Thickness and Hole Size Tolerance

1. .001 inch minimum copper, followed by .0005 inch tin-lead

Hole Tolerance	Drill Size to Use
±.002 (spread of .004 inch)	Nominal + .004 inch
+.003/−.002 (spread of .005 inch)	Nominal + .005 inch
±.003 (spread of .006 inch)	Nominal + .005 inch

2. .0015 inch minimum copper, followed by .0005 inch tin-lead

Hole Tolerance	Drill Size to Use
±.002 (spread of .004 inch)	Nominal + .005 inch
+.003/−.002 (spread of .005 inch)	Nominal + .005 inch
±.003 (spread of .006 inch)	Nominal + .006 inch

3. For .093-inch and thicker panels with a .001-inch minimum copper requirement, drill and plate as though .0015 inch minimum copper were the requirement.

Hole Tolerance	Drill Size to Use
±.002 (spread of .004 inch)	Nominal + .005 inch
+.003/−.002 (spread of .005 inch)	Nominal + .005 inch
±.003 (spread of .006 inch)	Nominal + .006 inch

to .093 inch, .125 inch, and greater. It is more difficult to achieve uniform plating thickness in high-aspect ratio holes. If the desired through-hole copper thickness is .001 inch, it may be necessary to plate .002 inch or more near the knee of the hole to obtain .001 inch minimum at the center of the hole (see Figure 3-2). It is common for planners and platers to allow a .001- to .002-inch greater hole size reduction in copper plating in order to be assured of meeting minimum through-hole thickness requirements. The tendency of high-aspect-ratio holes to plate thick near the knee of the hole and thin near the center is referred to as the "hourglass effect."

Table A-2 (MIL-STD-275E) recommends allowing an added tolerance spread of .004 inch for holes with aspect ratios of 4:1 or more. A hole with a diameter of .020 inch will have an aspect ratio of 3:1 in a .060-inch-thick board. A hole of .020 inch diameter in a board .120 inch thick will have an aspect ratio of 6:1. Thus, a .020-inch nominal plated-through hole will not always be drilled with the same size drill bit. The hole may be drilled out .001 to .002 inch or greater in a .120-inch panel than in a .060-inch panel. Of course, when holes in thick panels are drilled slightly larger, the plater must allow the hole to be reduced more in copper plating.

3. Size of the Panel. The size of the panel has an effect on the tolerances which can be held for much the same reason as aspect ratio: plating distribution. The greater the panel size, the greater the difference in plating rate between holes near the center of the panel and holes along the edges and

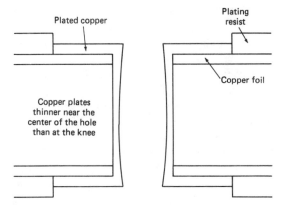

Fig. 3-2 The Aspect ratio and the Hourglass effect on plating: copper plates thinnest near the center of the hole and thickest at the knee.

corners. Current density and plating distributions are greatest at the corners and edges and least at the center of a panel (see Figure 3-3). The uneven distribution is of greatest concern for large panels with tight hole tolerances. An example of a large board or panel is 16 × 18 inches or greater.

4. Circuitry Layout. The circuitry layout refers to the distribution of metal areas. In general, the more even the pattern, the tighter the hole tolerance

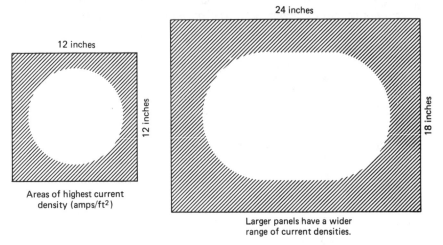

Fig. 3-3 Left: areas of highest current density. Right: larger panels have a wider range of current densities.

that can be held through plating. In the ideal situation, the holes and the circuitry are evenly distributed within the panel dimensions. Patterns which make it difficult to hold tight hole tolerances are the following:

a. Pads only on one side of the board, with circuitry and ground planes on the flip side. Patterns like this will cause the hole wall plating to become funneled; the holes will reduce quickly on the side with pads only, and more slowly on the side with circuitry. Planners must be alert for such jobs. This type of pattern is also referred to as "a single-sided board with plated-through holes, unless it is a multilayer.

b. Isolated holes. The circuitry may be fairly uniformly distributed, but may leave holes which have no circuitry near them. Such holes will plate much faster than holes which are near or surrounded by circuitry.

c. Isolated holes near the center of the board. In this special case, there is no circuitry in sections of the layout. Thus, it is possible to have dense circuitry over most of the panel, but isolated areas near the center of the pattern, with only a scattering of holes. Even though these holes are near the center of the panel, if there is no other metal areas near them, they are isolated and will plate much faster than holes elsewhere. Good design practice is to place metal around isolated areas in order to avoid hot spots with uneven plating. Some planners request permission to add this "nonfunctional" metal to the working film.

5. Annular Ring Requirements. The annular ring is the metal pad around the drilled hole. For external circuitry, the thickness of the metal plated inside the hole is considered part of the annular ring. For internal layers, the annular ring is measured from the actual drilled hole wall; plating thickness does not count (see Figure 3-4). The customer's annular ring requirement must be known in order to calculate whether any given drill size will allow for a adequate annular ring. It is sometimes necessary to use a smaller drill size than preferred so that annular ring requirements can be met.

   The most common requirements for annular rings are for the following:
   a. External
      (1) .005 inch (MIL-STD-275E and some corporations)
      (2) .002 inch (most common corporate requirement)
      (3) Tangency (allowed by some corporations)
      (4) Breakout (up to 25% of hole circumference allowed by some corporations)
   b. Internal
      (1) .002 inch (MIL-STD-275E and some corporations)
      (2) .001 inch (some corporations)
      (3) Tangency (some corporations)

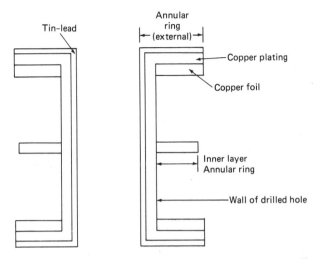

Fig. 3-4   Measurement of internal and external annular ring.

6. Inner Layer Clearance. This is the distance between the drilled hole and the inner layer power and ground planes. IPC-ML-910 Table IV C specifies this feature. It is treated as an air gap, which is the distance separating conductive patterns on a single layer. The only real requirement here is that the drilled hole not hit a power or ground layer or a trace on an internal signal layer. If this author were to establish written requirements for inner layer clearance, they would be as follows:

Where the greatest board dimension is less than 12 inches:

Preferred .020 inch or greater
Standard .015 inch
Reduced Producibility .010 inch

Where the greatest board dimension is 12 inches or greater:

Preferred .025 inch or greater
Standard .020 inch
Reduced Producibility .010 inch

Again, as with the annular ring, it may be necessary for the planner to call for a smaller drill size than would otherwise to preferred in order to achieve an inner layer clearance that provides a margin of confidence when all registration factors are taken into account.

7. Hole-to-Circuitry Air Gap. This is a special case similar to inner layer clearance. There are times when the circuit designer will allow the hole to be larger than the outer layer pad; this condition is called "wipeout." It is also possible that due to front-to-back misregistration, a drilled hole may break out of a pad and come closer to a trace or other conductive feature than is comfortable or allowed. In such cases, it is necessary to use the smallest practical drill size to avoid a short circuit or air gap problem. There are times when the printed circuit manufacturer must request permission to move or reroute a trace.

8. Front-to-Back and Layer-to-Layer Registration: How to Use It When Calculating Clearance and the Annular Ring. Registration must be discussed here because of its effect on the annular ring, inner layer clearance, and hole-to-trace air gap on outer layers, and when a wipeout occurs. Almost any degree of misregistration can be tolerated, provided it does not lead to an annular ring, inner layer clearance, or hole-to-trace airgap problem.

   To calculate annular ring and inner layer clearance, a table like Table 2 must be completed for each hole size (see Table 3-2 and Figure 3-5). The information in Table 3-2 is an integral part of incoming artwork inspection. It will be presented again in Chapter 4. Basically, from the pad or clearance diameter, the following are subtracted: drill size, misregistration, and shop tolerance (processing tolerance). The resulting value is then divided by 2: this will be the actual annular ring or clearance in the finished printed circuit, assumming everything is processed optimally.

   The shop tolerance is the best registration that any shop can hold, given the inaccuracies of artwork, imaging, drilling, programming, temperature, and humidity. Commonly used values for making this calculation are $\pm.002$ or $\pm.003$ inch. The planner must include a value for shop tolerance, since it is real and will affect registration, the annular ring, and clearance at the inner layers. By including a shop tolerance, the planner will obtain a much more realistic appraisal of how a given part number will proceed through the shop.

9. Drilling: Throughput and Cost. The drill size selected may also have an impact on the cost of drilling a given job and on the speed with which a job can be drilled. It is common for drillers to reduce the feed and speed for drill sizes below a certain diameter, usually about .030 to .025 inch. Panels which are .062 inch thick are typically drilled in stacks of three or

Table 3-2   Charts for Calculating Annular Ring and Inner Layer Clearance

ANNULAR RING

Hole Code	Pad Size	Drill Size	Mis-Registr.	Shop Tol.	Result	$\dfrac{\text{Result}}{2} = \dfrac{\text{Annular}}{\text{Ring}}$
____	____	____	____	____	____	_____
____	____	____	____	____	____	_____
____	____	____	____	____	____	_____
____	____	____	____	____	____	_____

INNER LAYER CLEARANCE

Hole Size	Clrnc. Dia.	Drill Size	Mis-Registr.	Shop Tol.	Result	$\dfrac{\text{Result}}{2} = \dfrac{\text{Inner}}{\text{Clearance}}$
____	____	____	____	____	____	_____
____	____	____	____	____	____	_____
____	____	____	____	____	____	_____
____	____	____	____	____	____	_____
____	____	____	____	____	____	_____

Instructions
1. Hole size: for each hole code, write in the nominal hole size.
2. Measure the pad or inner layer clearance for each hole size, and write this value under "Pad Size" or "Clearance."
3. Write in the diameter of the drill bit which will be used for each hole size.
4. Write in the amount of misregistration measured. If outer layer pads are being measured, this value will be the front-to-back misregistration. If inner layer clearances are being measured, or the inner layer annular ring, the value entered will be the worst-case inner layer misregistration measured.
5. Write in the shop tolerance currently being used. This will most likely be $\pm.003$ inch.
6. From the pad or clearance diameter, subtract drill size, misregistration, and shop tolerance. The value obtained is the result. This is the difference in diameters of the pad/clearance and the drill size after all registration factors have been subtracted.
7. Result/2 = annular ring or inner layer clearance − after all registration factors have been subtracted. This is the actual value likely to be measured on the final printed circuit board.

higher. However, when the drill diameter is small, the stack may be reduced to two or even one high. Obviously, this will have a serious impact on throughput. Reducing a drilling stack from three high to two or one high will double or triple the drilling time, the number of setups, and the number of drill bits required to drill the job. Once parameters are established for feed and speed, and for stack height as a function of the smallest drill size, the planner should, if possible, avoid calling out drill sizes which will have a negative impact on throughput. There are times when it will not hurt to go .001 or .002 inch over size, even if customer approval is required.

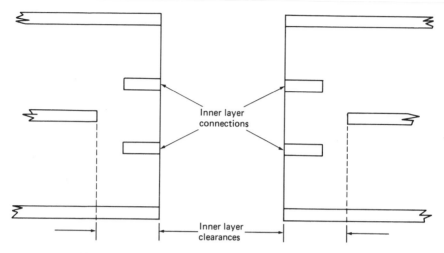

Fig. 3-5   Inner layer clearance at power and ground layers.

10. Finished Hole Size and Tolerance. Clearly, these factors are the most important consideration in choosing the proper drill bit for any given hole. Before the drill bit can be selected, some thought must be given to meeting the tolerance specified. And before the tolerance is assigned by the designer, some thought must be given to the actual tolerance required and on what the industry considers practical (i.e., IPC and military standards).

Tolerances for holes are listed as a function of three items: (a) hole size, (b) board thickness or aspect ratio, and (c) producibility.

a. Hole size. Generally, the larger the hole, the greater the tolerance which should be allowed. This is reflected in Tables B-7 (IPC-D-300G) and A-2 (MIL-STD-275E). Planners and designers should keep this in mind when assigning tolerances to holes.

b. Aspect ratio. This calculation allows the board thickness to be taken into consideration. The aspect ratio is defined or calculated as follows:

$$\text{Aspect ratio} = \frac{\text{board thickness}}{\text{hole diameter}}$$

The board thickness is divided by the diameter of the hole. In calculating the aspect ratio, the smallest plated hole size is used. Normally, the aspect ratio is not of great concern. The information in the tables listed above will result in aspect ratios of about 3 : 1 or less. The only time aspect ratio is of great concern is when it is greater than 3:1. This happens with very small holes and thick boards (such as .093 inch or greater). The planner should be most concerned when figuring drill sizes for thick boards. A hole diameter of .030 inch has an aspect ratio of about 2:1 for a .062-inch-thick board. A diameter of .030 inch in a .125-inch-thick board will result in an aspect ratio of about 5:1. Thin copper in the center of holes may or may not be of functional concern. However, it is of great concern for boards being built to MIL-P-55110. Boards being built to this specification are microsectioned at the coupons. The holes in the coupons are required to be drilled identical to the smallest plated holes of the circuit. The IPC and military standards recommend adding more tolerance for holes with aspect ratios greater than 3:1 or 4:1.

c. Producibility. It is difficult to build a printed circuit in which all the holes of a given hole code come out the same, unless it is a small circuit being processed one circuit per panel. The plating distribution over a panel surface will not be uniform. Holes at corners and isolated areas (areas with little metal nearby) will reduce more quickly than those near a panel center and those surrounded by a lot of metal circuitry.

Platers monitor hole size at various locations on a panel. A hole coded A, which is supposed to have a finished size of .040 ± .005 inch, is not a problem even for a mediocre plater. However, if the hole is to be .040 ± .002 inch, a much more skillful plater will be required. He/she will have to make sure that holes in the central area of the panel have adequate thickness, and that holes at corners and isolated areas do not reduce below the minimum and still have tolerance left for tin-lead plating. This is no easy task. If high aspect ratio holes, thick panels, and an uneven circuitry layout of the plating pattern are also factors, the task of meeting tight tolerances becomes enormous. This is why hole tolerances of ±.002 inch are listed as reduced producibility by MIL-STD-275E.

Planning and design personnel should note that IPC and military documents do not allow a ±.002-inch tolerance for holes larger than .063 inch or for high aspect ratio holes. Designers should think twice before putting a ±.002-inch tolerance on a blueprint. Planners should think twice before committing the printed circuit manufacturing shop to building boards with this tolerance, since it may have a negative impact on yield, on ability to meet delivery schedules (or to procure boards on time), and on costs. In addition, there is almost always a higher manufacturing scrap rate for tightly toleranced holes.

CONSIDERATIONS FOR NONPLATED HOLES

How to Make the Hole Nonplated

The simplicity of drilling a nonplated hole should not mislead the planner to overlook common pitfalls which cause scrap and rework. There are three basic methods for producing a nonplated hole for most printed circuits (single-sided boards will be discussed later):

1. Primary drill, along with the plated holes; then plug during pattern plating.
2. Primary drill; then tent over the nonplated holes with photoresist during the imaging operation.
3. Second drill at some operation after pattern plating.

Primary drilling is generally considered to be the most accurate method of locating a hole. All the holes are drilled with one setup on the drill, all are referenced from the same zero, and all holes on a given panel are drilled with the same spindle. However, there are only two reliable ways to keep the desired holes from plating: plugging and photoresist tenting. Each of these methods has its own drawbacks.

Hole plugging can be used to prevent plating in selected holes for boards imaged with photoresist or with screen printing. However, hole plugging is highly operator dependent. All of the desired nonplated holes must be plugged on each image of each panel. If the wrong hole is plugged, the most likely result will be a scrap board. It is fairly common for plugs to come out during handling or plating. This places the burden on the inspection department to catch the problem. Many platers have the habit of placing their fingers on the surface of the panel while plugging holes. This can result in fingerprints being plated into the circuitry. If the fingerprint falls on a contact finger which is to be nickel/gold plated, the result is scrap. The need to plug a large number of holes increases the likelihood of one or more wrong holes being plugged. Another point to consider is that this operation requires a lot of time for plugging and unplugging.

Photoresist tenting requires that dry film photoresist be the imaging method. Not all shops are set up for it, and not all shops use it exclusively. Photoresist is commonly considered to be a more expensive imaging technique than screen printing. Also, tenting does not work well for large (.080 inch or greater) holes. If the tent breaks, plating will take place in the hole. Exposure and developing control also play a part in the quality of the hole tent. If the hole tents break, the photoresist will flake off and may redeposit elsewhere on the plating surface of the panel. When this occurs, the redeposited resist flakes may create pitted plating and isolated areas of nonplating. This can be very costly in terms of yield loss and rework time.

Second drilling has the advantage of being relatively fast and accurate. It does require an extra operation, since the panels must be loaded back onto the drilling equipment. Because a new setup is involved (pinning the panels to the drilling table, running the N/C tape, and locating the correct zero), there is an opportunity to lose an extra .001 to .002 inch or more of registration. The added setup and the reduced drill accuracy are the chief disadvantages of second drilling.

In the absence of any other considerations, the decision to plug a few holes or second drill may boil down to how much time is available on the N/C drilling/routing equipment. Tenting should be undertaken when the shop has a clear idea of how large a hole the resist will reliably tent.

Second Drilling: When to Do It

Second drilling is often done just prior to fabrication, when the boards are routed from the panel with their final outline. It can even be done on a driller/router. One reason not to perform second drilling as part of fabrication is that it produces an effect called "drill shatter." The laminate will shatter and leave a damaged halo around the hole. If the second drilling is performed just prior to copper foil etching, for example, the copper foil will act as entry material and prevent drill-shattered holes. For some types of laminate material, especially Teflon and polyimide, second drilling after plating prior to etching copper foil is highly recommended. Teflon may not drill cleanly otherwise; it may take on a shredded, stringy appearance. Polyimide will have a yellow halo around all second-drilled holes if they are drilled after foil has been etched.

Hole Diameter Tolerance for Nonplated Holes

At first, the idea of hole tolerance for nonplated holes might seem to be not worth discussing. There are, however, some good reasons to think about the tolerance which is applied to nonplated holes. A planner or designer should consider the following before assigning a tolerance tighter than ±.002 inch to a nonplated hole:

1. The tightest tolerance which military literature requires is ±.002 inch. The IPC has the same requirements, except that they do allow ±.001 inch for Class C (their most severe class) on holes less than 0.33 inch. In general, a nonplated hole tolerance of less than ±.002 inch is not in keeping with industry standards.

2. A hole which is drilled at, say, .040 inch will almost never measure .040 inch; rather, it will usually measure .039 inch. If the hole is toleranced at

±.001 inch, the manufacturer is automatically at the bottom of the range. Thus, a ±.001 inch tolerance is really no tolerance at all.

3. Three common methods for measuring a hole are:
   a. Plug gauge—discrete pins of a specific diameter.
   b. Quick check gauge—a tapered shaft with calibrated markings.
   c. Microscope—usually a 60× or some other type of optical comparator.
   People who are experienced in measuring holes using these three methods know that they will often differ .001 inch or more from each other. Thus, if a hole is toleranced at ±.001 inch, this is really no tolerance at all, since there is no way of knowing how a hole will be measured by each method.

4. There are many drill bits available to the printed circuit manufacturer. However, if the manufacturer were to keep a production quantity of each size on hand at all times, a fortune would be tied up in drill bits. Most manufacturers keep a good selection of all commonly used sizes on hand, as well as a few of most—but not all—of the others. If there is sufficient tolerance on a hole, the manufacturer has greater freedom when choosing a drill size (See Table 3-3 for a listing of available drill sizes.) It is not practical to tolerance a hole at, say, .132 ± .001 inch; the drill sizes are just not available. It is practical to tolerance the hole .132 ± .003 inch, however. More drill sizes are available in the ±.003-inch range. As with any tolerance, a designer who ignores industry standards does as much of a disservice to his/her company as a planner who does not question the tolerance which does not meet industry standards.

Metal Pads/Annular Rings Around Nonplated Holes

Not all nonplated holes have metal pads around them. If at all possible, it is good to leave them off or remove them during photo processing of working film. Table B-1 of IPC-D-300G lists annular ring requirements for nonplated holes (unsupported holes). The smallest, Class C, is .015 inch. Annular rings on nonplated holes present problems. The metal will burr when the hole is drilled. This creates an undesirable condition and may result in time-consuming rework. Also, the metal may lift during the drilling operation and may have to be picked off anyway. If a metal pad is desired around the nonplated hole, it should have a relief inside so that the drill will hit only laminate, not metal. The pads will require a donut shape, not a solid dot, on the working film. The clear area inside the donut should be about .005 inch greater in diameter than the drill size.

Table 3-3 Drill Sizes

Sizes	Decimal Inches	Sizes	Decimal Inches	Sizes	Decimal Inches	Sizes	Decimal Inches	Sizes	Decimal Inches	Sizes	Decimal Inches
97	.0059	59	.0410	2.75 mm	.1083	5 mm	.1969	7.6 mm	.2992	1/2	.5000
96	.0063	1.05 mm	.0413	7/64	.1094	8	.1990	N	.3020	13 mm	.5118
95	.0067	58	.0420	35	.1100	5.1 mm	.2008	7.7 mm	.3031	33/64	.5156
94	.0071	57	.0430	2.8 mm	.1102	7	.2010	7.75 mm	.3051	17/32	.5312
93	.0075	1.1 mm	.0433	34	.1110	13/64	.2031	7.8 mm	.3071	13.5 mm	.5315
92	.0079	1.15 mm	.0453	33	.1130	6	.2040	7.9 mm	.3110	35/64	.5469
.2 mm	.0079	56	.0465	2.9 mm	.1142	5.2 mm	.2047	5/16	.3125	14 mm	.5512
91	.0083	3/64	.0469	32	.1160	5	.2055	8 mm	.3150	9/16	.5625
90	.0087	1.2 mm	.0472	3 mm	.1181	5.25 mm	.2067	O	.3160	14.5 mm	.5709
.22 mm	.0087	1.25 mm	.0492	31	.1200	5.3 mm	.2087	8.1 mm	.3189	37/64	.5781
89	.0091	1.3 mm	.0512	3.1 mm	.1220	4	.2090	8.2 mm	.3228	15 mm	.5906
88	.0095	55	.0520	1/8	.1250	5.4 mm	.2126	P	.3230	19/32	.5938
.25 mm	.0098	1.35 mm	.0531	3.2 mm	.1260	3	.2130	8.25 mm	.3248	39/64	.6094
87	.0100	54	.0550	3.25 mm	.1280	5.5 mm	.2165	8.3 mm	.3268	15.5 mm	.6102
86	.0105	1.4 mm	.0551	30	.1285	7/32	.2188	21/64	.3281	5/8	.6250
85	.0110	1.45 mm	.0571	3.3 mm	.1299	5.6 mm	.2205	8.4 mm	.3307	16 mm	.6299
.28 mm	.0110	1.5 mm	.0591	3.4 mm	.1339	2	.2210	Q	.3320	41/64	.6406
84	.0115	53	.0595	29	.1360	5.7 mm	.2244	8.5 mm	.3346	16.5 mm	.6496
.3 mm	.0118	1.55 mm	.0610	3.5 mm	.1378	5.75 mm	.2264	8.6 mm	.3386	21/32	.6562
83	.0120	1/16	.0625	28	.1405	1	.2280	R	.3390	17 mm	.6693
82	.0125	1.6 mm	.0630	9/64	.1406	5.8 mm	.2283	8.7 mm	.3425	43/64	.6719
.32 mm	.0126	52	.0635	3.6 mm	.1417	5.9 mm	.2323	11/32	.3438	11/16	.6875
81	.0130	1.65 mm	.0650	27	.1440	A	.2340	8.75 mm	.3445	17.5 mm	.6890
80	.0135	1.7 mm	.0669	3.7 mm	.1457	15/64	.2344	8.8 mm	.3465	45/64	.7031
.35 mm	.0138	51	.0670	26	.1470	6 mm	.2362	S	.3480	18 mm	.7087
79	.0145	1.75 mm	.0689	3.75 mm	.1476	B	.2380	8.9 mm	.3504	23/32	.7188
1/64	.0156	50	.0700	25	.1495	6.1 mm	.2402	9 mm	.3543	18.5 mm	.7283
.4 mm	.0157	1.8 mm	.0709	3.8 mm	.1496	C	.2420	T	.3580	47/64	.7344
78	.0160	1.85 mm	.0728	24	.1520	6.2 mm	.2441	9.1 mm	5.383	19 mm	.7480

Table 3-3 (Continued)

Sizes	Decimal Inches	Sizes	Decimal Inches	Sizes	Decimal Inches	Sizes	Decimal Inches	Sizes	Decimal Inches	Sizes	Decimal Inches
.45 mm	.0177	49	.0730	3.9 mm	.1535	D	.2460	23/64	.3594	3/4	.7500
77	.0180	1.9 mm	.0748	23	.1540	6.25 mm	.2461	9.2 mm	.3622	49/64	.7656
.5 mm	.0197	48	.0760	5/32	.1562	6.3 mm	.2480	9.25 mm	.3642	19.5 mm	.7677
76	.0200	1.95 mm	.0768	22	.1570	E	.2500	9.3 mm	.3661	25/32	.7812
75	.0210	5/64	.0781	4 mm	.1575	1/4	.2500	U	.3680	20 mm	.7874
.55 mm	.0217	47	.0785	21	.1590	6.4 mm	.2520	9.4 mm	.3701	51/64	.7969
74	.0225	2 mm	.0787	20	.1610	6.5 mm	.2559	9.5 mm	.3740	20.5 mm	.8071
.6 mm	.0236	2.05 mm	.0807	4.1 mm	.1614	F	.2570	3/8	.3750	13/16	.8125
73	.0240	46	.0810	4.2 mm	.1654	6.6 mm	.2598	V	.3770	21 mm	.8268
72	.0250	45	.0820	19	.1660	G	.2610	9.6 mm	.3780	53/64	.8281
.65 mm	.0256	2.1 mm	.0827	4.25 mm	.1673	6.7 mm	.2638	9.7 mm	.3819	27/32	.8438
71	.0260	2.15 mm	.0846	4.3 mm	.1693	17/64	.2656	9.75 mm	.3839	21.5 mm	.8465
.7 mm	.0276	44	.0860	18	.1695	6.75 mm	.2657	9.8 mm	.3858	55/64	.8594
70	.0280	2.2 mm	.0866	11/64	.1719	H	.2660	W	.3860	22 mm	.8661
69	.0292	2.25 mm	.0886	17	.1730	6.8 mm	.2677	9.9 mm	.3898	7/8	.8750
.75 mm	.0295	43	.0890	4.4 mm	.1732	6.9 mm	.2717	25/64	.3906	22.5 mm	.8858
68	.0310	2.3 mm	.0906	16	.1770	I	.2720	10 mm	.3937	57/64	.8906
1/32	.0312	2.35 mm	.0925	4.5 mm	.1772	7 mm	.2756	X	.3970	23 mm	.9055
.8 mm	.0315	42	.0935	15	.1800	J	.2770	Y	.4040	29/32	.9062
67	.0320	3/32	.0938	4.6 mm	.1811	7.1 mm	.2795	13/32	.4062	59/64	.9219
66	.0330	2.4 mm	.0945	14	.1820	K	.2810	Z	.4130	23.5 mm	.9252
.85 mm	.0335	41	.0960	13	.1850	9/32	.2812	10.5 mm	.4134	15/16	.9375
65	.0350	2.45 mm	.0965	4.7 mm	.1850	7.2 mm	.2835	27/64	.4219	24 mm	.9449
.9 mm	.0354	40	.0980	4.75 mm	.1870	7.25 mm	.2854	11 mm	.4331	61/64	.9531
64	.0360	2.5 mm	.0984	3/16	.1875	7.3 mm	.2874	7/16	.4375	24.5 mm	.9646
63	.0370	39	.0995	4.8 mm	.1890	L	.2900	11.5 mm	.4528	31/32	.9688
.95 mm	.0374	38	.1015	12	.1890	7.4 mm	.2913	29/64	.4531	25 mm	.9843
62	.0380	2.6 mm	.1024	11	.1910	M	.2950	15/32	.4688	63/64	.9844
61	.0390	37	.1040	4.9 mm	.1929	7.5 mm	.2953	12 mm	.4724	1	1.0000
1 mm	.0394	2.7 mm	.1063	10	.1935	19/64	.2969	31/64	.4844		
60	.0400	36	.1065	9	.1960			12.5 mm	.4921		

Drilling Single-Sided Panels

Single-sided boards have their own set of problems. These include: (1) hole reduction during plating, (2) metal rings to be picked off, and (3) drilling from the wrong side of the copper cladding.

1. Hole Reduction during Plating. How can the size of nonplated holes be reduced during plating? This is not as silly as it might seem. Hole size reduction is a common problem when a hole with a metal pad or annular ring is drilled. At the lip of the hole is the copper foil. The thickness of the foil forms part of the hole wall. While metal will not plate at the laminate part of the hole wall, it will plate at the foil part (see Figure 3-6). It is only the lip of the hole that is actually reduced. The only way to prevent this reduction is to reduce the amount of plating on the surface of the board. Since the foil of the laminate forms the conductor, there is little need for copper electroplating anyway. A few minutes of flash plating will suffice. Should thicker conductors be required, 2 oz foil can be used in place of 1 oz.

2. Metal Rings to Be Picked Off. What are metal rings, and why must they be picked off? Since all the holes in a single-sided board are nonplated, there

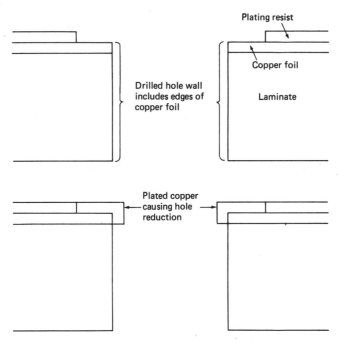

Fig. 3-6 Hole size reduction on a single-sided board due to plating.

is no reason for second drilling. Right? Wrong. There are two categories of metal rings: those on the metal side of a single-sided board and those which form on the back (blank) side. Metal rings tend to form on ᵔingle-sided boards at holes which are not intended to have a metal pad. Metal rings also form on the blank side of single-sided boards for the same reason. If double-sided laminate is being used to make a single-sided board, the copper should be etched from one side of the panel prior to drilling (see Figure 3-7). The same phenomenon which causes size reduction in nonplated holes will cause unwanted metal rings to form. When the hole is to have no annular ring or metal pad, the plating resist comes right up to the lip of the hole. However, it does not go into the hole and cover the foil part of the hole wall. The foil part of the hole wall will plate with copper and then with tin-lead. After etching, tin-lead protects the copper foil part of the hole wall. All copper around the hole is etched away, except for the metal ring which has formed at the lip of the hole wall.

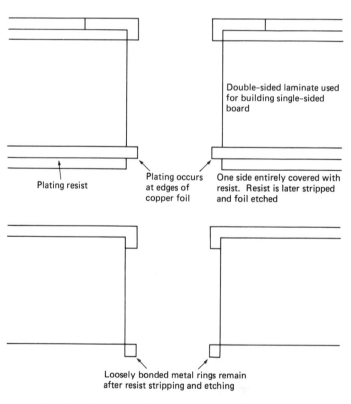

Fig. 3-7 The cause of metal rings on single-sided boards.

These metal rings are totally unwanted by the printed circuit manufacturer and by the customer. The manufacturer must either spend a lot of time picking them from the holes or must ignore them and ship the boards with the rings. Is there an alternative? Yes. Metal rings on the metal side of a single-sided board can be prevented by second-drilling all holes which are not intended to have an annular ring. These padless holes are simply drilled after the plating operation. If second drilling is performed prior to etching copper foil, any possibility of drill shatter can be avoided. Obviously, metal rings can be prevented on the back side by simply etching foil from that side before drilling. There is, however, another practice which will help prevent the metal ring problem on both sides of the board. The metal rings are thin and very loosely adherent to the laminate after etching. If plating is kept to a minimum, say a 5- or 10-minute copper flash prior to tin-lead plating, the rings will not form enough to present a problem. When the plating is thin, the rings will be removed by the etching operation. It is not good practice to start with double-sided laminate for single-sided boards unless the laminate is etched prior to drilling. However, this leads to the third and last problem.

3. Drilling from the Wrong Side of the Copper Cladding. Drilling from the wrong side of the copper cladding can be a problem. The drill operator must know whether to place the foil up or down when loading a stack of single-sided panels on a drill table. There is a 50-50 chance of building the board backward. The last thing that is needed is a working film that will not align to the side which has copper. The blueprint should be marked "this side metal," and the drill program's first article should be marked "drill with this side up." It is a good practice for a printed circuit manufacturer to establish the procedure of drilling all single-sided jobs with the copper side either up or down. The planner, programmer, and drill operators should agree on this procedure.

HOLE LOCATION TOLERANCES

There are realistic tolerances which must be assigned in determining the accuracy with which a drilled hole can be located in a printed circuit board. Generally, an N/C drill can locate holes within a tolerance of $\pm.001$ inch of its programmed location. However, laminate will change dimensionally as panels are processed; the material will shrink or grow. Therefore, the tolerances for locating drilled holes in the finished product must be looser than the location tolerances of the N/C drilling equipment. Also, when a drill manufacturer states that a given machine is capable of, say, $\pm.0005$ inch this accuracy applies to the movement of the drill table. There is another factor: the collet runout of the individual spindles. The drill spindle contributes to a loss of drilling accuracy. Many N/C drill manufacturers recommend that spindles be replaced when collet runout exceeds .0005 to .001 inch. The collet runout is a measure of how accurately the drill turns on its center. Worn collets, worn bearings, dragging a drill across a panel, physical abuse, and other factors add to excessive drill collet runout.

The IPC and military documents publish standards for hole location tolerances. Refer to Table B-8 (IPC-D-300G) and to Table A-2 (MIL-STD-275E) for these values. There are two cases which depend on whether the greatest dimension is greater or less than 12 inches. There are also three classes: A, B, and C (IPC) and Preferred, Standard, and Reduced Producibility (military). Note that the tightest rtp is .002 inch. Blueprints which are printed with dimensioned hole location tolerances of $\pm.002$ inch may look nice on paper, but they are not really sound.

BOARD OUTLINE TOLERANCES

This is commonly referred to as "fabrication tolerances." They apply when a board is being routed from the panel. This is a requirement which MIL-STD-275E does not address as well as it should. IPC-D-300G does address it and lists tolerances for classes A, B, and C (see IPC-D-300G Table 13-7). Note that class C, the most severe tolerance, is listed as $\pm.006$ inch. If MIL-STD-275E did address this item, class C would correspond to the military's Reduced Producibility class. The $\pm.006$ inch of tolerance is perhaps a little generous for the printed circuit manufacturer. A three-place tolerance (.XXX on the blueprint) is typically specified as $\pm.005$ inch. Manufacturers using N/C routers can almost always meet this requirement. Pin routing, which uses a prefabricated outline of the board, will cause difficulty in achieving $\pm.005$ inch. The $\pm.005$-inch tolerance is generally considered to be the limit of epoxy/fiberglass or related laminate material. Tolerances tighter than $\pm.005$ inch should always be avoided. If looser tolerances will cause no problem, they should be specified on the blueprint.

CONDUCTIVE PATTERN LOCATION

This is the location tolerance for such things as surface mount pads, contact fingers, and land areas when referenced from a datum or board edge. Table B-9 (IPC-D-300G) and Table A-2 (MIL-STD-275E) list the tolerances for these items. There are three classes based on degree of difficulty (or manufacturability), and a further distinction is made according to whether the longest dimension is greater or less than 12 inches.

Sometimes blueprints reference dimensions from the centerline of a contact finger. This is an invalid dimensioning method for tightly toleranced requirements. Any tolerance quoted must be equal to or greater than the values listed in the above table plus the board outline tolerance. Again, it is fairly common to see tolerances of $\pm.005$ inch for the centerline of contact finger-to-board edges. But such tolerances are not in keeping with the reality of printed circuit manufacturing and industry standards. This type of tolerancing should be questioned by the planner.

ACHIEVING CIRCUIT CONFORMANCE TO THE ARTWORK

This section deals with what the printed circuit manufacturer must do to ensure that the circuitry conforms to the artwork within the allowed tolerances. It is common for procurement specifications to state that no modifications may be made to the artwork without written approval of the procuring entity. The only recommendations to modify artwork that will be made here are those which are necessary to allow for normal processing tolerances. IPC and military specifications discuss the need to compensate artwork for processing tolerances.

Considerations which the planner must be sensitive to include the following:

1. Trace width and minimum trace width
2. Minimum conductor spacing (air gap)
   a. Trace to trace
   b. Trace to pad
   c. Pad to pad
3. Hole-to-conductor spacing
   a. Inner and outer layers
   b. Inner layer clearance
4. Annular ring
5. Layer-to-layer (or front-to-back) registration
6. Solder mask and legend considerations
7. Imaging choice:
   a. Dry film
   b. Screen printing
8. Image to be made for the plating pattern or etching pattern
   a. For plating patterns, will final metal be tin-lead or a nonreflowable metal?
9. Electrical testing requirements

Factors Affecting Trace Width for Plating and Etching Patterns

The IPC and military specifications cover trace width tolerances. Both have three classes of producibility. The classes are, as always, A, B, and C for the

IPC and Preferred, Standard, and Reduced Producibility for the military. Table B-11 of IPC-D-300G and Table A-2 of MIL-STD-275E list the trace width tolerances. These tables list the tolerance as functions of foil thickness and plated or nonplated circuitry. Note that MIL-STD-275E has slightly more difficult requirements than the IPC specification. Many of the class C (Reduced Producibility) parameters discussed thus far can be achieved by diligent operators at each of the processes involved and by allowing high overage to compensate for reduced yield. Meeting tight trace width tolerances, however, can be a different story. The copper foil cladding of the laminate determines how much trace reduction will result after etching. Refer to Figure D-3 for a diagram of how foil thickness and etching affect trace width. There are various rules of thumb which describe the relationship between foil thickness and trace width reduction through etching; for example, for every .001 inch of foil thickness there will be a .001-inch trace width reduction from etching. Table II of IPC-D-310A even lists how much to compensate the artwork for various processes. This table is not reproduced in this book. The information in the IPC-D-310A standard can be somewhat simplified. Table 3-4 lists recommendations based on the author's experience. Below is a list of factors which can affect finished the conductor width for printed circuit boards. Since it is not the purpose of this book to deal with process control, it will be assumed that all of these considerations are under normal good process control.

1. Trace width on the artwork
2. Exposure control
   a. Artwork
   b. Dry film photoresist
   c. Stencil for screen printing
3. Plating pattern versus etching pattern
4. Screen printed plating pattern versus dry film photoresist plating pattern
5. Amount of plating
6. Thickness of base foil for plating and etching patterns

Table 3-4   Compensating Incoming Film in Order to Generate Working Film for Process Tolerances

1. Dry film: Plating or etching pattern[1]
   1 oz copper foil: Spread trace width .002 inch.
   2 oz copper foil: Spread trace width .004 inch.
2. Screen-printed plating pattern[2]
   1 or 2 oz copper foil: Choke trace width .002 to .004 inch.

[1]*Note:* Trace widths on working film must be greater by the above values than the incoming customer's film to compensate for the reduction which will occur during processing.
[2]*Note:* Trace widths on the working film must be reduced by the above values, over the incoming customer's film, to compensate for the growth which will occur during processing.

7. Type of etchant used
8. Skill of the etching operator
9. Type of metallic etch resist

Plating Patterns: Dry Film Image versus Screen-Printed Image.
Generally, conductors which are plated with a screen-printed image tend to be wider than they are on the artwork. Conductors plated from a dry film photoresist image tend to be narrower than they are on the artwork. That is, finished traces will grow wider with screen-printed plating patterns and reduce with dry film plating patterns (see Figures 3-8 and 3-9). Generally, the trace width will increase .002 to .003 inch after plating when the image has been screen printed. For dry film photoresist image plating patterns, trace widths will reduce .002 inch on 1-oz foil cladding and reduce .004 inch on 2-oz foil cladding. The loss in conductor width is due to etching.

Dry film resist forms fairly straight channels for plating. Futhermore, the channels are as deep as the resist is thick, typically .0015 to .002 inch. Plating is confined to the channels formed by the resist. Unless the plating is grossly excessive, it will not result in an increase in conductor width. The plated trace width is very close to that of the dry film image and to the artwork. After plating, the resist is stripped from the panel. The copper sides of the plated conductors are exposed to the etchant. The copper sides of the conductors tend to reduce during the etching process. For this reason, the conductor width of the working film (often referred to as the "production master") must be expanded, or spread, to compensate for etching losses. The etching losses are proportional to the thickness of the copper foil cladding of the laminate. The

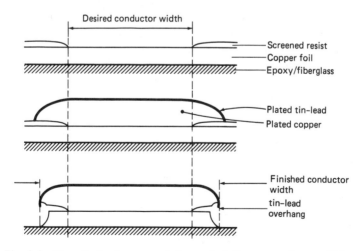

Fig. 3-8 The effects of screen printing and plating on conductor width.

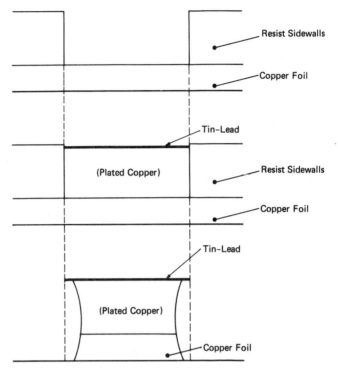

Fig. 3-9   The effect of photoresist imaging and plating on conductor width.

thickness of the copper plating on top of the foil has little effect on etching losses. There is another factor to consider with dry film exposure. The final trace width can be manipulated approximately .0005 to .001 inch by slightly underexposing or overexposing the resist. Some printed circuit manufacturers underexpose routinely. For this reason, it sometimes appears that etching reduces the trace width .001 to .002 inch on 1-oz foil instead of a full .002 inch. It may also appear that traces are reduced .002 to .003 inch on 2-oz foil instead of a full .004 inch.

Screen-printed resist is thinner than dry film, typically less than .001 inch. The channels do not have straight sides; instead, they slope outward. The width will start to increase almost immediately with plating. The trace bulges out over the resist during plating. This overplating also keeps the trace width from reducing significantly during etching. Even fairly heavy overetching will not result in trace width reduction for a circuit which has been imaged by screen printing. It is the plating over the resist which causes conductor growth and creates the need to compensate the artwork. Conductor width on the working film (production master) must be reduced to compensate for the growth which will occur. Stencil preparation also causes the trace width to grow from .001 to .002 inch.

The trace width increase can become a serious problem, especially where conductor spacing on the artwork is minimal, say, less than .008 to .010 inch. It should also be clear than for every .001 inch of trace width gained, the conductor spacing (air gap) will decrease by .001 inch. For every .001 inch of trace width lost, the conductor spacing will increase by .001 inch. Conductor spacing and trace width are inversely related.

Practical Limits on Screen Printing and Dry Film Photoresist. Screen printing is less exacting for printed circuit requirements than photoresist imaging. It is difficult to reliably screen print conductor widths less than .008 inch across. The difficulty is due to the surface tension of the ink being printed, off-contact distance between the screen and the panel surface, stencil thickness, touchup difficulties, and the high solvent content of the ink itself. If conductor width increases .002 to .003 inch with plating and stencil preparation, conductor spacing (air gap) will decrease .002 to .003 inch. Artwork which has an air gap (conductor spacing) of .008 inch will result in an air gap of .006 to .005 inch on the printed circuit pattern after plating. Utilizing screen printing for such an application may create problems in resist stripping and etching, second touchup, presolder mask inspection, solder masking, and electrical testing. The author recommends using dry film photoresist for imaging when the conductor width and spacing are tighter than .008 inch. It should also be noted that some screeners are capable of printing successfully in the .005- to .006-inch trace width range. However, this requires great skill and process control on the part of the screen printer. Screen printing .005- to .006-inch trace widths and spaces should never be considered a routine production application.

Dry film photoresist can be used reliably to achieve conductor widths and spacing of .005 inch. The chief difficulty when imaging in this range is touchup. Artwork from the customer must be of sufficient quality that little or no touchup is needed. Since plating is confined to channels formed by the resist, increases in condutor width are of no concern. (See the next section on apparent effects of non-tin/lead etch resists.)

Summary: Screen-Printed Imaging. Trace widths on the working film (production master) should be decreased (choked) a minimum of .002 inch less than the master film supplied by the customer. If the annular ring and trace width will allow more choke, this should be performed. A practical limit is reached when the conductor width has been reduced to about .008 inch. This compensates for growth due to plating and stencil preparation.

Summary: Dry Film Imaging. Trace widths on the working film (production master) should be increased (spread) .001 to .002 inch for 1 oz cladding, and increased .002 to .004 inch for 2 oz cladding, above the master film supplied by the customer. This compensates for losses due to etching.

NOTE: There are instances where etching loss will reduce the trace width by .002 to .004 inch but no compensation to the working film is desired. An example of this is the need to increase an air gap (conductor spacing). If the .002 to .004 inch of trace width reduction is within the limits allowed by the customer, there is no need to compensate the artwork.

Non-Tin/Lead Etch Resists: Apparent Effect on Trace Width. One of the advantages of tin-lead as an etch resist is that it is a reflowable metal; it melts fairly easily. During the etching operation after copper and tin-lead plating, there is tin-lead overhang. The overhang is caused by copper being etched out just beneath the tin-lead along the conductor edges. After reflow, the overhang melts and covers the sides of the conductors. This happens only when the overhang is tin-lead. This sometimes causes people to feel that reflow reduces trace width. What actually happens is that the tin-lead melts, eliminating the overhang, so that it becomes apparent that the trace width has been reduced. If the overhang is tin, nickel, or tin-nickel, the overhang will remain after reflow. The copper conductor width has actually been reduced, but it is not apparent since only the tin, nickel, or tin-nickel is visible (see Figure 3-10).

Etching Pattern: Considerations. The chief consideration in regard to etching pattern is conductor width. Most etching patterns are used for inner layers on multilayer printed circuitry. Inner layer etching patterns are almost always applied with dry film photoresist. In practical terms, traces will reduce about .001 to .002 inch below artwork when 1 oz copper foil is used and .002 to .004 inch when 2 oz copper foil is used. If the goal is to obtain finished trace widths which are virtually identical to those on incoming master artwork (the customer's film), the working film must be compensated. In this case, the trace width on the working film must be spread .002 inch (1 oz copper foil) or .004 inch (2 oz copper foil) to compensate for etching losses. If the .002 to .004 inch of trace reduction is within the limits allowed by the customer, there is no need to compensate the artwork.

Power and Ground Inner Layers. The chief problem with regard to internal power and ground layers is inadequate clearance for the drilled holes where no electrical connection ot the internal layer is desired. When a hole is to go through the board without making contact to a layer, a clearance must be made in that layer for the hole. The clearance must be large enough to accommodate:

1. The drilled hole
2. Allowances for artwork misregistration

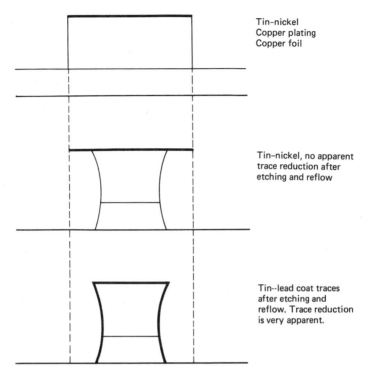

Tin–nickel
Copper plating
Copper foil

Tin–nickel, no apparent
trace reduction after
etching and reflow

Tin--lead coat traces
after etching and
reflow. Trace reduction
is very apparent.

Fig. 3-10   Non-tin/lead metallic etch resists (such as nickel and tin-nickel) mask actual trace reduction from etching.

3. Allowances for layer-to-layer misregistration
4. Allowances for drilling misregistration
5. Specified conductor spacing requirements
6. Normal dimensional changes in the laminate due to processing, especially thermal processing

We have already discussed how to calculate inner layer clearances, and we will deal with this again in Chapter 4. There are times when it is determined that the inner clearance provided for on the incoming artwork is inadequate. This means that there is too great a possibility of drilling into the metal of the ground or power plane. It is necessary to increase the clearance by performing a photographic spread. It is generally permissible to spread the clearance to any convenient size.

The most common pitfalls when enlarging the inner layer clearances are:

1. Reducing tie bar connections at thermally relieved holes.

2. Creating electrically isolated holes where a connection to the internal plane is desired.
3. Reducing the trace width on internal plane layers which also have conductor traces.
4. Reducing the annular ring at inner layer connections below a specified minimum.

These problems will now be discussed in greater detail.

1. Thermally relieved holes make connections to the internal plane. There are relieved areas around the holes to enhance solderability. Connection is made through these relieved areas by the tie bar (see Figure 3-11). If the diameter of the clearance is increased, say, .010 inch, the width of the tie bar will be reduced .010 inch. If the tie bar width (trace width) is .015 inch on the incoming artwork, it will be only .005 inch after the clearance diameter is increased by .010 inch. Therefore, it is necessary to measure the width of the tie bars when determining how much to increase a clearance. A decision must also be made on the minimum tie bar width that the shop is willing to accept. The starting and ending tie bar widths normally place limits on increases in clearance diameters. It should never be forgotten that the tie bar width will decrease about .001 to .002 inch for 1-oz copper and .003 to .004 inch for 2-oz copper because of etching; clearance diameter will change by approximately the same amount. If the tie bars are destroyed through etching or a photographic process, the resulting printed circuit will have electrically open circuits.

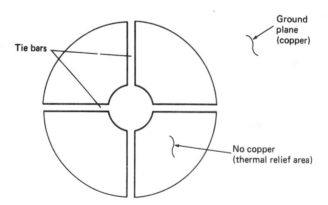

Fig. 3-11 Diagram of a thermally relieved connection to an internal plane where the connection is made by a plated-through hole.

2. Electrical isolations or open circuits can accidentally be created by increasing clearance diameters, as discussed above. However, increasing clearance diameters too much causes another problem (see Figure 3-12). The artwork inspector and planner must pay attention to the distance (amount of metal) separating rows of clearance and thermal relief holes in power and ground planes. This distance must be measured and known before determining how much to spread a clearance. As the clearance diameter is increased, the distance between clearances decreases. Wiping out the metal between clearance holes is not necessarily a problem. It becomes a problem only if the thermally relieved connections are surrounded by rows of clearances and the metal connecting the tie bars to the plane is wiped out.

3. Some power and ground planes contain conductor traces. It must be kept in mind that increasing a clearance by .010 inch in diameter will also reduce the trace width by .010 inch. Minimum trace width and trace width tolerances must be taken into consideration before issuing instructions on increasing clearance diameters.

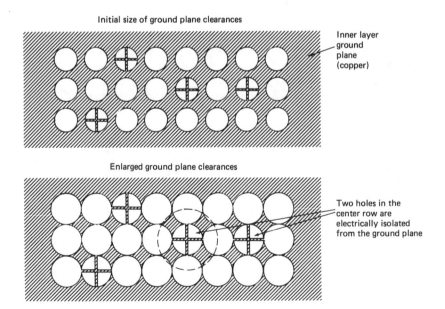

Initial size of ground plane clearances

Inner layer ground plane (copper)

Enlarged ground plane clearances

Two holes in the center row are electrically isolated from the ground plane

Fig. 3-12 Example of how photographically increasing clearances can create holes which will be electrically isolated from internal power and ground planes.

4. When conductive traces are present on internal plane layers, there must be some consideration given to the specified internal annular ring. Every .010 inch of clearance diameter increase will reduce the internal annular ring by .005 inch.

Solder Mask Requirements

Solder masking is probably the most difficult aspect of building printed circuit boards today. This is chiefly due to the tight trace-to-pad spacing. The requirements of surface mount technology, and especially, solder mask over bare copper (SMOBC), have added greatly to solder masking difficulties. Tight pad-to-trace spacing makes the application of solder mask without getting mask on pads or exposing traces difficult. Surface mount and SMOBC mean that solder mask on pads will result in scrap product. However, if the planner understands the limitations in applying solder mask, he/she can help set up difficult jobs with as few problems as possible. Also, there are several solder mask resists on the market, each with markedly different properties. The facts about solder mask which planners and designers should be aware of include the following:

1. Few customers prefer solder mask on the pads of plated holes. However, almost all of them will accept solder mask on the pads as long as it is not inside the holes and as long as the job is not SMOBC. Some customers stipulate that the mask should be .002 inch or more from the lip of the hole. This provides a .002-inch annular ring free of solder mask. Solder mask is generally acceptable when it is on pads because it does not cause a soldering problem. Solder will flow easily beneath solder mask. Solder mask on pads is generally not acceptable when a job is being run SMOBC. This is because it is aesthetically unappealing, giving the appearance of exposed copper on the pads.

2. Most customers do not want solder mask on surface mounting pads. However, most are beginning to allow solder mask to cover part of the pads. Generally, the mask must not cover more than 20% of the pad's surface area along one edge, with a limit of about .004 inch in from the edge. If the mask is encroaching on the pad from all directions, or if there are isolated globules on the interior portion of the pad, the printed circuit board manufacturer will have a serious acceptability problem.

3. Most customers do not want traces to be exposed by the opening for pads. It is generally preferred that traces be entirely encapsulated by solder mask resist. Many customers believe that there is no reason for traces to be exposed anyway, since they allow solder mask on the pads – even surface mounting pads, to some extent.

4. Given the choice between solder mask on pads and exposed traces, most customers prefer mask on the pads.

5. In order to guarantee no solder mask on pads, at holes, or at surface mounting pads, there must be at least .010 inch of clearance between the edge of the solder mask opening and the pad on the circuitry. This assumes optimum registration in drilling, imaging, and solder masking. Optimum registration means zero front-to-back misregistration on the incoming customer film and holding all other registration factors in the manufacturing shop to ±.003 inch (see Figure 3-13).

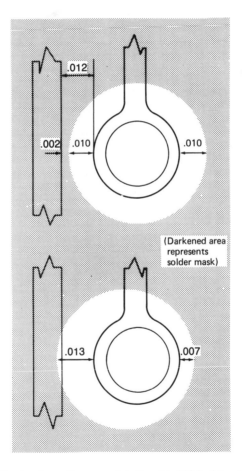

Fig. 3-13 Even allowing for optimum registration (±.003 inch), a pad-to-trace air gap of .013 inch is required to guarantee no mask on pads and no exposed traces.

Another way to state this clearance requirement is that the solder mask opening must be at least .020 inch greater in diameter than the corresponding pad on the circuitry film.

For every ±.001 inch of added misregistration (shop tolerance) the shop allows, another .002 inch must be added to the opening diameter of the solder mask. For every ±.001 inch of front-to-back misregistration, another .002 inch must be added.

6. Trace-to-pad spacing must be at least .013 inch to guarantee that the traces will not be exposed by the solder mask openings at the pads. This is true of traces near round pads of holes and square surface mounting pads. Under the best circumstances, the solder mask will register within ±.003 inch of the pad. Figure 3-13 shows that when the solder mask is registered perfectly to the circuitry, there is .010 inch of clearance all the way around the pad. When the solder mask is registered within ±.003 inch, it will be .007 inch away from the pad on one side and .013 inch away on the other side. If a trace is less than .013 inch to the pad, it will be partially exposed by the solder mask opening. The trace must be at least .013 inch from the pad to assure complete coverage.

The requirement for a .013-inch pad-to-trace air gap explains why solder mask application is the most difficult task in building printed circuits today. Current state-of-the-art circuitry has .008-inch traces and conductor spacing. In order to screen print solder mask, the mask must be allowed on the pads. The most difficult type of board to solder mask (not counting SMOBC) is the type that contains surface mount pads with traces running between the pads. In such cases, pad-to-trace air gaps are .006 to .008 inch. Rarely are air gaps more than .008 inch.

7. Surface mount boards with traces running between pads often have pad-to-trace air gaps of .006 to .008 inch. It is virtually impossible to produce this type of circuit using SMOBC unless at least one of the following circumstances occurs:
   a. Dry film solder mask is being used.
   b. A waiver for exposed traces is obtained.
   c. Mask on the surface mount pads is allowed. This means that copper will be visible on the surface mount pads after solder application.

8. Numerous types of solder mask materials are available. They fall into four categories:
   a. Dry film solder mask
   b. Ultraviolet (u.v.) curable, screen-printed solder mask

c. One- or two-part epoxy screen-printed solder mask

d. Non-dry-film, photo-imagable solder mask

a. Dry film solder masks will be mentioned only briefly. These are good-looking masks with several attractive features. They are applied in a vacuum laminator and imaged with artwork and u.v. light, just like dry film. They can easily deal with .008-inch pad-to-trace air gaps. However, they are prohibitively expensive for many customers. The setup cost is about $200,000, which is steep for most captive or job shops. As a service available from a number of sources, this setup will add $6 to $8 per square foot to the cost of the printed circuit, which is prohibitively expensive for many customers. This added expense drives up the cost of the printed circuit and makes U.S. manufacturers less competitive with foreign competition.

b. The u.v. curable solder masks have excellent screening properties. They do not bleed and are virtually 100% solids. This means that they can be screen printed for tighter tolerance work than high-solvent, heavy-bleeding materials common to epoxy masks. They offer the printed circuit manufacturer a definite advantage for tight (less than .013-inch spacing) work.

c. Epoxy solder masks have been available for a long time. They are the most commonly used type of mask and are screen printed. These masks tend to be fairly high in solvent. Typically they are 50% to 70% solids, the remainder being solvent. The solvent must be driven off in an oven. The high solvent level contributes to a problem known as "bleed," whereby the uncured resist, being led by evaporating solvent, creeps outward from the area where it has been applied by screen printing. The bleed causes mask to creep onto surface mount pads and pads around holes. This may or may not cause scrap for traditional reflowed solder boards. However, for solder mask over bare copper applications, the bleed gives the appearance of exposed copper after solder has been applied by hot air leveling. The hot air–leveled solder will not coat copper which has solder mask bleed on it. This results in an aesthetically unappealing appearance, which is almost always unacceptable to customers buying printed circuits. Solder mask bleed is a serious problem to printed circuit manufacturers and poses a continual threat to high yield for SMOBC and SMT circuits.

d. Non-dry-film, photo-imagable solder masks are fairly new and not widely used. They have great potential for solving many of today's solder mask problems. Basically, the entire panel is coated with the resist. Artwork is registered to the panel. U.V. light exposes the resist according to the artwork. The unexposed resist is then developed from the panel surface. Ideally, the mask will cover traces with no bleed onto pads, and as accurately as dry film solder mask.

9. The screening holes used to register the panels for solder mask application must be nonplated. These holes are primary drilled. Typically, they are plugged during pattern plating. Photoresist tents may hold and should be attempted. If these holes are not plugged, they will plate through, forcing the screening operator to pin to plated holes. For several reasons, most of which have been discussed, not all holes of a given drill size will have the same finished size after plating. They will differ on the same panel, on separate panels of the same job, and from one job to another; they will even differ from plater to plater.

Screening holes are typically drilled at .126 inch (3.20-mm drill bit). The screening registration pins are typically .125 inch. This provides a snug fit for good, consistent registration. Without a snug fit, registration will differ from panel to panel. When screen operators are forced to pin to plated holes, they must file their pins to fit inside the holes. Filed pins are never accurate. They range from around .110 to .123 inch from pin to pin and on the same pin, if measurements are taken at 90 degrees on the pin.

10. Screen printing has several inherent misregistration factors:
    a. The off-contact distance between the stencil and the panel surface (see Figure 3-14). The squeegie presses the stencil-bearing screen to the panel surface, creating a slight angle between them. The greater the off-contact distance, the greater the angle and the resulting misregistration.
    b. Dimensional changes in the screen mesh. The screen mesh is made from

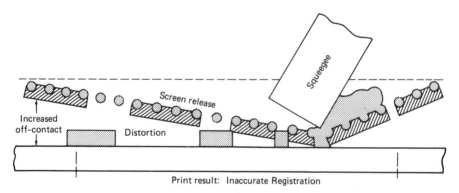

Fig. 3-14 Effect of high screen mesh off-contact distance on registration. (from Clark, *Handbook of Printed Circuit Manufacturing.*)

polyester fibers, which must be continually retensioned by the screening operator. Also, the pressure of the squeegie will distort the mesh (and stencil).

c. The stencil is an intermediate imaging medium between the working film and the solder mask image.

11. What can the planner do to overcome solder mask difficulties? Fortunately, several courses of action can be taken. Some examples are as follows:

a. Tight pad-to-trace air gaps are a common problem. If the trace width is decreased, the air gap will increase. This air gap increase (trace width decrease) can be achieved by photographically modifying the film or by using dry film imaging and allowing normal etching losses to take their toll. The main concern is that the final circuit matches the artwork within specified tolerances. Many customers, as well as MIL-P-55110, specify $\pm 20\%$ change as being acceptable. This means that a pad-to-trace air gap can be increased 20% and still meet customer specifications.

NOTE: Jobs with tight pad-to-trace air gaps should not be imaged using screen printing technology, which results in trace widths increases and air gap decreases. This situation will only make solder mask application more difficult.

b. Tightly toleranced jobs should be run on smaller panels. An alternative is to set jobs up for shearing just prior to solder masking if the panel has multiple images. It is easier to achieve and hold tight registration on smaller panels.

c. Solder masks which bleed minimally, or not at all should be specified. U.V. curable, and several of the one- and two-part solder masks:

d. Tightly toleranced and SMOBC jobs should have instructions on the traveler to blot repeatedly between each panel. This keeps the stencil dry and prevents resist from being smeared on surfaces where it is not desired. These types of jobs require a minimum of three blotting pushes with the squeegie between panels.

e. Keep surface plating as thin as possible, while still meeting the customer's specifications. Plating which is very high off the surface is difficult to screen over, creating dry areas and smearing the resist.

f. If the pad is wide – e.g., if it has an annular ring of .006 inch or more – there is little difficulty in allowing mask on the pad, as long as there is about .002 to .003 inch of mask-free solder at the lips of the hole. If the annular ring is less than .006 inch, a potential problem exists. Printed circuits which are to be electrically tested cannot have solder mask covering the pad up to the hole; there must be at least .002 inch of annular ring around the lip of the hole which is free of solder mask. If the combination of pad-to-trace air gap and allowable mask on pads does not leave a .002-inch minimum annular ring, the planner cannot release the job to the floor. Alternatively, he/she must consider dry film or another photo-imageable solder mask – which the customer must be asked to pay for.

g. The planner should involve the screen printing supervisor before releasing the job. Reviewing the requirements and difficulties will help the job flow through the solder mask area. The planner can communicate the customer's expectations to the supervisor. If a job is unusually difficult, there is nothing wrong with instructing the supervisor. A brief review with the supervisor prior to releasing the job will allow the planner to take advantage of the supervisor's comments.

Screening operators often do not know what is most important on a given job. If certain areas must be covered by solder mask (such as traces between surface mounting pads), there should be a note on the traveler to inform them. If mask on via hole pads is acceptable, there should be a note stating so. Special notes will allow the screener to devote attention to the areas which are most important.

Legend Requirements

The legend is the information screened onto the panels to help the customer identify certain areas and components. It is often referred to by the customer as "silk screen." There are limitations and conditions regarding the legend which the planner must be aware of. Problems associated with legends generally fall into the following categories:

1. Character width is less than .007 inch. Characters which are made from line widths less than .007 inch do not print clearly; they will break up and only partially print. The result is illegible characters. The character width must be photographically spread to be at least .007 inch.

2. The characters fall directly over pads, holes, or surface mount pads. If the customer will allow epoxy legend ink on pads, there is no problem. To prevent ink from getting into holes, the planner must instruct the photo department to "blow the hole centers out." This is a technique that photo technicians use to remove dots in the center of holes on incoming artwork as part of the process for producing working film. Even though there are no hole centers to be removed on legend film, this same technique will remove the legend characters where they align directly over holes.

3. Registration cannot be absolutely determined. Most legend film has lines connecting holes or boxes and circles surrounding holes. In such situations, there is no question about how the film is to be registered to the panel. However, sometimes the legend film is composed only of numbers, symbols, or other characters. This makes it difficult to determine exactly how the film is to be registered. The planner must recognize this potential problem and issue instructions on how to deal with it. One way to assure that screen printing operators know how to register the artwork is to add lines or circles connecting or surrounding holes. This can be done in a specified and agreed-upon location, so that the screening operators will recognize these lines or circles as a registration aid.

MULTILAYER CONSTRUCTION

In multilayer construction, there are some elementary factors which all planners should be aware of. This knowledge will allow the shop to procure the least expensive materials, use the least expensive constructions, and complete the job in the least amount of time with the fewest problems. The goal in this section is not to produce an expert in multilayer manufacturing, but rather to produce planning engineers who are aware of factors which can cause unnecessary scrap and cost for the manufacturer.

Board Thickness Tolerance

IPC and military requirements for board thickness are identical. MIL-STD-275E and IPC-ML-910 call for board thicknesses of $+10\%$ of nominal thickness or .007 inch, whichever is greater. Any requirement on a blueprint or procurement specification which is tighter than this should be recognized as being more stringent than the industry standard.

Common Laminates and Prepregs

1. A variety of laminate thicknesses are available to the board manufacturer. It is common for laminate manufacturing companies to list more sizes than they actually have on hand. Thus, any callout by the planner should reflect what can actually be made available in a short time. Also, any printed circuit shop must restrict itself to a variety of standard thicknesses. This reduces the number of laminates which must be inventoried. Commonly inventoried laminate thicknesses are listed in Table 3-5.

2. Copper foil is referred to in ounces. For example, 1-oz foil weighs about 1 oz per square foot and is about .0014 inch thick. Common foils are:
   a. $\frac{1}{2}$ oz = .00072 inch

Table 3-5   Commonly Inventoried Multilayer Laminates

.005 1/1
.008 1/1
.010 1/1
.013 or .014 1/1
.022 1/1
.028 or .031 1/1 (virtually identical)

NOTE: 2/2 or 2/1 foils are less common. Check with vendors for availability prior to committing the shop to a production schedule. NOTE: Half/half (H/H) foils are becoming more popular. Check with vendors for availability.

   b. 1 oz = .0014 inch

   c. 2 oz = .0028 inch

   An "H," for half, is often used in place of $\frac{1}{2}$ for half-ounce foil. Laminate thickness followed by $\frac{1}{1}$ means 1-oz foil on both sides; $\frac{2}{2}$ means 2-oz foil on both sides; $\frac{2}{1}$ means 2-oz foil on one side and 1-oz foil on the other.

3. Laminates which are .0309 inch or thinner are referred to by core thickness; the number assigned to the thickness applies to the core thickness only, and does not include copper foil cladding. Thus .022 $\frac{1}{1}$ means .022-inch-thick core material with 1-oz copper foil on both sides. The actual laminate would measure about .022 + .0028 = .0248 inch.

4. Laminates which are .031 inch or thicker are referred to by the combined thickness of core plus copper. Thus, .059 $\frac{1}{1}$ laminate would measure about .059 inch, including the copper foil; .028 $\frac{1}{1}$ would measure about .031 inch; .031 $\frac{1}{1}$ would measure about .031 inch. Thus, .028 $\frac{1}{1}$ and .031 $\frac{1}{1}$ are almost identical materials. Laminate which is .031 inch or thicker must contain a water mark in the core. The water mark identifies the laminate manufacturer. The only difference between .028 $\frac{1}{1}$ and .031 $\frac{1}{1}$ is the presence of a water mark in the .031 $\frac{1}{1}$ material.

5. Prepreg is uncured epoxy resin coated over fiberglass cloth. It is available in a variety of thicknesses. The thickness is usually determined by the style of the fiberglass cloth. Prepregs are generally referred to by their glass cloth style. Thus, 104, 108 or 1080, 112 or 2112, 113, 116 or 2116, and 7628 are all common glass (or prepreg) styles. Generally, the thicker the glass, the lower the resin content. Prepreg thickness is usually referred to as the

average pressed thickness, and is very close to the thickness of the glass used to make it (see Table 3-5).

6. Prepregs in the range listed in Table 3-6 are about the same price, regardless of thickness. The 7628 prepreg is actually about 5% cheaper per square foot than the 112 or 2112. This is because 7628 contains less epcxy resin. Thus, four sheets of 2116 would have an average pressed thickness of .014 inch. Two sheets of 7628 prepreg would have the same thickness, but would cost less than half as much. Thicker prepregs should be used over thinner ones to build large dielectrics.

7. Never use only one sheet of prepreg. This is forbidden by MIL-P-55110 and MIL-STD-275. When deciding how to construct a given printed circuit, try to use two sheets of prepreg. Go to three sheets if needed, but never use only one. Military specifications also require that double-sided laminates contain a minimum of two sheets of prepreg. For military applications, .007 $\frac{1}{1}$ double-sided laminate, must be built from two sheets of prepreg. Most laminate manufacturers use one sheet of 7628 between the sheets of foil. The planner must make sure that the purchasing department knows how to specify material for military applications.

8. Laminates which are .0309 inch or thinner are about the same in price, varying by about 10%. Using thicker laminate for a given construction eliminates the need to use prepreg to create dielectric thickness. Using less prepreg not only keeps the cost down but improves dimensional stability during lamination. Thick laminate is easier to process than thin laminate. It can be scrubbed, dried, etched, developed, and handled with less chance of becoming damaged. This is another reason to use the thickest laminate that is practical when deciding on a construction.

Table 3-6　Average Pressed
Thickness of One Sheet of Prepreg

Prepreg Style	Average Pressed Thickness (Inch)
104	.0015
108 or 1080	.002
112 or 2112	.003
113	.0035
116 or 2116	.004
7628	.0065

9. Avoid the use of "caps" which are single-sided outer layers (see Figure 3-15). Caps use an extra sheet of laminate and two extra sheets of prepreg. This adds about $3 per square foot to the cost of a panel. If a capping method of construction is desired, the shop should consider the use of foil caps.

10. When deciding how to construct any given multilayer, build for a pressed thickness near the bottom of the tolerance range. This is less expensive and leaves more room for thickness which will be added during plating. Thus, a board which is to be .062 + .007 inch can be built as thin as .055 inch coming out of the press. About .005 inch of thickness will be added during plating. This will bring the finished board thickness close to the nominal thickness without running the risk of being too thin or too thick.

11. Copper foil has an effect on the final thickness of any panel. Generally, there are two types of inner layers: signal layers (traces) and power or ground planes. The copper foil of power and ground planes contributes to the thickness of the panel; the copper foil of the signal layers does not. The traces press into the prepreg and become completely encapsulated. Because the copper traces are "buried" in the prepreg, their thickness does not contribute to the overall thickness. Planners should beware of 2-oz copper for internal signal layers. Two-ounce signal layers will encapsulate, reducing the dielectric spacing between adjacent layers by almost .003 inch. An

Standard construction for four-layer board

.022 1/1		Cost	2.00/ft²	× 1	= $ 2.00
Two sheets 2112		Cost	.38/ft²	× 2	= .76
.022 1/1		Cost	2.00/ft²		= 2.00
		Total material cost			= $ 4.76/ft²

Cap construction for four-layer board

.010 1/0		Cost	1.95/ft²	× 1	= $ 1.95
Two sheets 2112		Cost	.38/ft²	× 2	= .76
.022 1/1		Cost	2.00/ft²	× 1	= 2.00
Two sheets 2112		Cost	.38/ft²	× 2	= .76
.010 0/1		Cost	1.95/ft²	× 1	= 1.95
		Total material cost			= $ 7.42/ft²

Conclusion: Cap construction for a four-layer board costs $2.66/ft² more than standard construction. This is more than a 50% greater materials cost.

Fig. 3-15 Cost comparison: two constructions for a four-layer multilayer printed circuit.

added sheet of prepreg may be required to meet minimum dielectric spacing requirements.

When the planner is reviewing a set of blueprints, he/she should look at the layer sequence chart and note whether each internal layer is signal or plane (see Figure 3-16). This chart will be referred to by others as well.

12. Watch out for minimum dielectric callouts on a blueprint or in a specification. They can create cost and manufacturability problems.

    a. MIL-P-55110 and MIL-STD-275 call for a minimum dielectric spacing between layers of .0035 inch. This in itself is not a problem, but it is necessary to review copper foil requirements and signal versus plane layer sequencing. Two sheets of 2112 prepreg will result in a dielectric spacing between layers of about .006 inch when the layers are planes

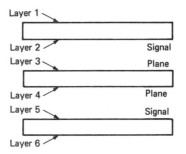

Fig. 3-16 Layers denoted as signal (traces) or plane.

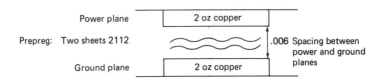

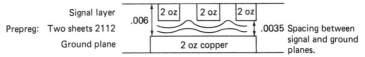

Fig. 3-17 Signal traces encapsulate in prepreg, reducing dielectric spacing. Power and ground layers do not encapsulate. The full dielectric potential of prepreg is achieved when prepreg separates internal planes.

(see Figure 3-17). However, two sheets of 2112 prepreg will result in a dielectric spacing of only about .0035 inch when separating a 2-oz signal layer from a ground plane.

b. Minimum dielectric requirements can create overall thickness problems and result in the need for more prepreg. Figure 3-18 shows the effect

1. Thickness .059±.006, standard construction: no minimum dielectric spacing between layers 2 and 3

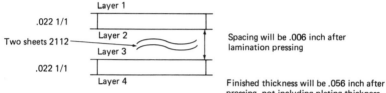

Layer 1

.022 1/1

Two sheets 2112 — Layer 2 / Layer 3

Spacing will be .006 inch after lamination pressing

.022 1/1

Layer 4

Finished thickness will be .056 inch after pressing, not including plating thickness

2. Thickness .059±.006, standard construction: .012 inch minimum spacing between layer 2 and 3 (7628 prepreg to be used)

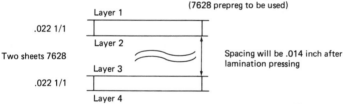

Layer 1

.022 1/1

Layer 2

Two sheets 7628

Layer 3

Spacing will be .014 inch after lamination pressing

.022 1/1

Layer 4

Possibility that postplating thickness will exceed .059±.006

Finished thickness will be .062 inch after pressing, not including plating thickness

3. Thickness .059±.006, cap construction: .012 inch minimum spacing

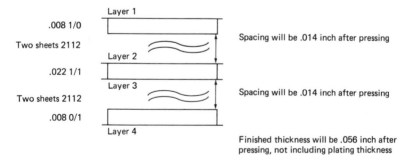

Layer 1

.008 1/0

Two sheets 2112

Layer 2

Spacing will be .014 inch after pressing

.022 1/1

Layer 3

Two sheets 2112

Spacing will be .014 inch after pressing

.008 0/1

Layer 4

Finished thickness will be .056 inch after pressing, not including plating thickness

Fig. 3-18 Effect of a minimum dielectric spacing requirement on a four-layer printed circuit.

that a minimum dielectric callout can have on board construction. If the callout will affect the construction choice, it may be best to discuss the requirement with the customer.

13. Multilayer boards should be run with at least one coupon per panel. The coupon can be used to microsection a panel in order to measure dielectric

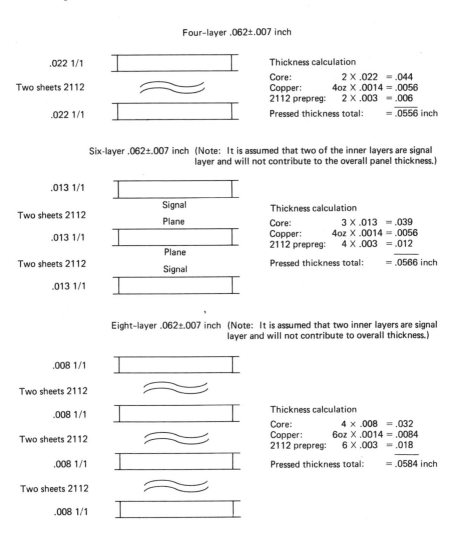

Four-layer .062±.007 inch

.022 1/1	Thickness calculation
Two sheets 2112	Core: 2 X .022 = .044
.022 1/1	Copper: 4oz X .0014 = .0056
	2112 prepreg: 2 X .003 = .006
	Pressed thickness total: = .0556 inch

Six-layer .062±.007 inch (Note: It is assumed that two of the inner layers are signal layer and will not contribute to the overall panel thickness.)

.013 1/1
Signal
Two sheets 2112
Plane
Thickness calculation
.013 1/1
Core: 3 X .013 = .039
Plane
Copper: 4oz X .0014 = .0056
Two sheets 2112
2112 prepreg: 4 X .003 = .012
Signal
Pressed thickness total: = .0566 inch
.013 1/1

Eight-layer .062±.007 inch (Note: It is assumed that two inner layers are signal layer and will not contribute to overall thickness.)

.008 1/1
Two sheets 2112
.008 1/1
Thickness calculation
Two sheets 2112
Core: 4 × .008 = .032
Copper: 6oz X .0014 = .0084
.008 1/1
2112 prepreg: 6 X .003 = .018
Two sheets 2112
Pressed thickness total: = .0584 inch
.008 1/1

Fig. 3-19 Standard constructions for four-layer, six-layer, and eight-layer printed circuits. The thickest core laminate is used, and two sheets of prepreg are required for each bonding location.

spacing, foil thickness, lamination, or through-hole quality at any time, without having to destroy a good board. The coupon should reflect the construction. Signal layers should have a signal-type design for the coupon on that layer; Plane layers should have a planar-type coupon. Thus, the coupon will accurately reflect the construction of the panel.

14. Track the direction of the grain of the laminate. The grain direction is also referred to as the "long cloth," "strong cloth," and "warp direction." Laminate is made from prepreg. Prepreg is made from fiberglass, which is woven and wrapped around a round core. The long direction is the grain, strong cloth, or warp direction. The short direction is the weft direction. Laminate expands at different rates during multilayer pressing. The warp, or grain, directions should always align with each other. If they cross at 90 degree angles, the laminated multilayer panels will have a severe warp or bow.

Two conveniently used sheet sizes are 48 × 36 and 48 × 72 inches. Some laminate manufacturers laminate only 48 × 72 sheets, which can be cut down to 48 × 36. Some manufacturers laminate both 48 × 36 and 48 × 72 sheets. It is important to know what sheet size the laminate vendor uses. A 48 × 36 laminated sheet will have the grain direction in the 48-inch dimension. A 48 × 72 laminated sheet will have the grain in the 72-inch direction. If the 48 × 72 sheet is cut into two 48 × 36 sheets, these sheets will have the grain in the 36-inch direction. Thus, it is possible for a printed circuit shop to receive 48 × 36 sheets with grain running in different directions. All laminate purchased should always have the grain in the same direction. It is a good precaution to have the laminate manufacturer place grain direction identifying arrows on each box or sheet of laminate.

15. Figure 3-19 lists examples of constructions for laminating four-, six-, and eight-layer printed circuits. The thickest practical laminate has been called out.

Chapter 4

Artwork Inspection

INCOMING INSPECTION

The first duty of the incoming inspector is to perform artwork inspection. Other duties include reviewing the blueprint, purchase order, and traveler for discrepancies. Artwork inspection must be a combination of visual assessment and physical measurement. The artwork inspector must be able to identify both clear and less obvious artwork and design errors. Less obvious errors generally require the inspector to take physical measurements and perform basic calculations. Incoming inspection is one of the first opportunities the printed circuit manufacturer has to demonstrate his/her efficiency. Customers depend heavily on their board vendor to catch errors. If the board manufacturer fails to do so in a timely manner, he/she has missed an opportunity to service the customer properly. By waiting until a job is underway to detect errors, the board manufacturer may have jeopardized the competitive position of the customer, and, ultimately, of himself. Poor or nonexistent incoming inspection creates poor customer–vendor relations in another way. It is easy to go to the customer with a design or manufacturability problem during the first day or two. Nothing causes greater customer–vendor problems than waiting until the job is ready to ship and then trying to get a waiver for a condition that could have been identified and handled when the job was being started. The printed circuit manufacturer must identify and solve manufacturability problems as soon as possible. A manufacturer who is not ready, willing, and able to do this may not remain competitive.

Figure 4-1 shows an example of basic minimum information which must be obtained during artwork inspection. The incoming inspector should complete the incoming inspection report. The report should have been started by the planner, who should also have done a preliminary artwork review. Each of the blocks in the incoming report will be covered separately below. Planning and inspection personnel should always keep in mind one very important rule: Do not make assumptions. It is better to verify than to assume. It is the responsi-

INCOMING ARTWORK INSPECTION REPORT

CUSTOMER	P/N	REV.	REPEAT	REMAKE
DATE INSPECTED	DRAWING NO.	REV.	NEW	CHANGE
INSPECTOR	ARTWORK NO.	REV.	NEW REV.	

NEGS	POS	S/M	LEGEND	PM	PRINTS	NO. LAYERS

			AIR GAP		
	REGISTRATION	MIN. TRACE	PAD/PAD	PAD/TRACE	TRACE/TRACE
LAYER 1					
LAYER 2					
LAYER 3					
LAYER 4					
LAYER 5					
LAYER 6					
LAYER 7					
LAYER 8					
OTHER					
S/M : COMPONENT : CLEARANCE					
: SOLDER : CLEARANCE					
LEGEND: COMPONENT					
: SOLDER					

Hole Code	Size	Tol.	P/NP	Pad Size	Drill Size	I/L	A/R	S/M
1)								
2)								
3)								
4)								
5)								
6)								
7)								
8)								
9)								

Fig. 4-1 Incoming artwork inspection report.

bility of the incoming inspector to catch discrepancies which occurred because someone else made an assumption.

Below are the steps involved in artwork inspection. They follow the sequence laid out in the incoming artwork inspection report.

Inner layer power to ground shorts

Pads hitting traces

Blueprint: Side showing identified _
: Artwork/drawing circuitry/hole patterns match. _ _ _ _ _ _ _ _ _ _ _
: Letters/numbers read same as artwork _ _ _ _ _ _ _ _ _ _ _ _ _ _ _
: Symmetrical hole pattern _ _ _ _ _ _ _ _ _ _ _ _ _ _ _ _ _ _
: Layer sequence clearly identified_ _ _ _ _ _ _ _ _ _ _ _ _ _ _ _ _

Artwork: Missing or extra pads _
: Fuzzy or ragged circuitry _ _ _ _ _ _ _ _ _ _ _ _ _ _ _ _ _ _ _
: Crossed traces or broken traces_ _ _ _ _ _ _ _ _ _ _ _ _ _ _ _ _
: Traces with nothing at end_ _ _ _ _ _ _ _ _ _ _ _ _ _ _ _ _ _ _
: Layers numbered correctly _ _ _ _ _ _ _ _ _ _ _ _ _ _ _ _ _ _ _

Notes:_ _

Dispositions _

QUALITY ASSURANCE MANAGER

Fig. 4-1 (*Continued*)

BASIC STEPS OF ARTWORK INSPECTION

1. Customer. The name of the customer is entered here. The purpose of this item may seem so obvious that it is not worth discussing. However, this is not the case. The customer's name is not necessarily the name on the blue-

print or artwork. As part of a turnkey operation, one company may design and/or procure printed circuit boards for another company. It is important that the actual customer be identified and that the incoming inspector and planner know exactly which company's specification to follow. In cases where one company is procuring printed circuits for another, the customer listed on the traveler should be the company that issued the purchase order. If the printed circuits will eventually be shipped to the company whose name is on the blueprint, it is permissible to have two names on the traveler. However, it should be clear who issued the purchase order. The second company, the one subcontracting to your customer, can have its name in parentheses. If the name of this second company is not on the traveler, there is the possibility of not being able to trace a blueprint or artwork to the traveler.

It is possible that two specifications have to be met and that one specification may conflict with the other. A specification summary sheet should be available to the incoming inspector. This is a one-page summary of important information from the customer's procurement specification. This sheet should be reviewed before continuing the artwork review. When two specifications have to be met, they must be reviewed for conflicting requirements. This is a fairly common situation. The incoming inspector should question the areas of conflict and list them on the incoming inspection report. This provides a record of the discrepancy and its ultimate disposition. There are important requirements which the incoming inspector must be aware of in order to perform an adequate inspection.

2. Part Number and Revision Level. This is divided into three sections on the incoming inspection report:
   a. Part number_____ Revision_____
   b. Drawing number_____ Revision_____
   c. Artwork number_____ Revision_____

   a. The part number is the number on the purchase order. The revision level of the part number is the revision level on the purchase order for that part number. If the purchase order is not available, the sales order form may serve as a substitute. The part number and revision level listed on the traveler and incoming inspection report should be the numbers listed on the order form submitted by the sales department.
   b. The drawing number and revision level should be the actual numbers listed on the blueprint (drawing). These must be identical to the numbers listed on the purchase order or sales order form. If there is more than one set of blueprints, all sets must be identical. If there is more than one sheet in a set of blueprints, all sheets must have identical part and revision numbers.

c. The artwork number and revision level should be identical to the numbers listed on the blueprint and purchase order. Often, however, the artwork number differs from the numbers on the other documents. When this happens, a note on the blueprint should reference the correct artwork and revision numbers to be used. These must agree with the artwork. Each piece of film provided by the customer must be checked. All pieces should have the same part and revision level.

NOTE: It is fairly common for each piece of film to have a different part number. The last digit of the part number may change sequentially. If the part numbers on the artwork differ more than this, the difference should be questioned.

NOTE: Sometimes two part numbers are assigned to artwork. One number, which is inside the fab lines of the circuit, is identical to the part and revision level on the purchase order and blueprint. The other number may be specified on the blueprint as the number which should be used. This second number may or may not be found inside the fab lines and is actually like a tooling number used internally by the customer. The tooling number for the artwork may have meaning only to the designer. This is the number which should be listed on the traveler and the incoming inspection report for artwork number. If this tooling number is not found inside the fab lines of the artwork, this must be noted on the blueprint or on the incoming inspection report. Unless a note is made, someone in the photo, imaging, or inspection department may question the apparent discrepancy.

The part number and revision level are divided into three sections on the traveler and the incoming inspection report because the customer will often submit drawings and artwork which differ from each other and from the part number and revision level on the purchase order. If the planner notices a discrepancy between any of these numbers, he/she must review the notes on the purchase order or blueprint for verification that the differing numbers or revision levels are supposed to be as submitted. If the documentation does not verify that the differing numbers are correct, the planner must stop further planning and resolve this discrepancy. If the planner has failed to notice the discrepancy, the incoming inspector must verify its correctness. If there is nothing to show that the discrepancy is intended, the incoming inspector must question it on the incoming inspection report and then return the entire package to the planning department for resolution.

NOTE: Always use the full part number. If the part number is 40-33224-002, this is the number which should be on all of the documentation. Some shops will delete, for example, the 40- or the -002. This is poor practice in any shop. All of these numbers have significance. In shops which do not insist on full part numbers for their documentation, it is fairly common to build a wrong revision or part number.

NOTE: Part numbers and revision levels on the purchase order, blueprint, and artwork should agree with each other. Any discrepancy should be investigated further.

NOTE: Do not proceed with planning or incoming inspection on a job with discrepant part or revision numbers. Remember: an incomplete part number is a discrepancy.

3. Job Status. Job status will fall into one of the following categories: new, repeat, new revision, remake, or change. Normally, only new jobs and new revisions will come to the incoming inspector for a report. If changes are being made internally at the request of the customer, these changes constitute a revision, and the job should go to the incoming inspector. New jobs are just that—new. The printed circuit manufacturer will have no previous documentation on the part number. It is the job of the planner to verify that a job submitted as new is, in fact, new. It is common to find that a job listed as new is actually a new revision. New revisions pose the problem of whether or not to try to use any of the prior tooling: artwork, drill program, router program, and electrical test fixture. Some companies make a practice of purging previous tooling and starting a fresh. Others have the incoming inspector review the new and old revisions for changes— the goal being, of course, to avoid performing programming and photo work if at all possible. Whether or not to attempt to use prior tooling is a matter of company policy.

Attempting to use tooling for previous revisions of a part number is always risky. However, it is a fairly common practice. If the incoming inspector is assigned the task of reviewing previous tooling, the following are some of the items to be checked:
a. Board outline: compare all dimensions on the blueprints for both revisions.
b. Hole charts: look for changes in hole coding, sizes, and the number of holes for each coding.
c. Hole changes
   (1) Check the blueprint to see if changes have been made in the location of holes. The number and size of B holes, for example, may not have changed. However, the B hole locations may have been altered.
   (2) Have changes been made regarding plating or nonplating of holes?
   (3) Have changes been made in the diameter tolerances?
d. Notes on the blueprint may give some indication of what changes were made. Study these notes.
e. Each layer of artwork, including legend and solder mask film, must be checked for artwork changes.

f. It should be determined whether the number or sequence of layers has been changed.

g. Customer comments. Sometimes the customer will verbally state the nature of the change(s) between the revisions. The incoming inspector and planner should consider these comments but should perform a thorough comparison for other changes. The printed circuit manufacturer should not be misled by statements from the customer. The manufacturer is still responsible if the customer erred and failed to mention all changes.

If there are no apparent changes in the drilling or routing programs, the incoming inspector should review the information with the programming supervisor. If the supervisor concurs, the drill/router tape boxes and file cards should be changed to reflect the new revision. This should be done immediately and verified by the incoming inspector. The revision level on the drill first article and mylar should also be changed, dated, and signed by the incoming inspector. The same situation exists in regard to artwork. If the incoming inspector is unable to find any differences, the artwork should be reviewed by the photo supervisor.

4. Catalogue the Artwork and Blueprints. In the boxes provided on the incoming inspection report, list what was actually received from the customer.
    a. How many negatives?
    b. How many positives?
    c. How many solder mask films?
    d. How many legend films?
    e. How many pad masters?
    f. How many blueprints?
    g. How many layers to the printed circuit?

This is something which should already have been filled out on the incoming inspection report as part of the planner's review of the job. Virtually the entire top section of the incoming inspection report should have been filled out by the planner. Knowing this information is part of doing a careful and complete job of planning. This artwork review and cataloguing helps prevent numerous errors which might otherwise go unnoticed or waste a lot of time. For instance:

a. There may be two separate legends, one for the component side and one for the solder side. Unless there is a note confirming this fact on the blueprint, it may go unnoticed until the customer wants to return the boards. If there are two separate legend films, the traveler must call for a legend on both sides.

b. There may be two separate solder mask films because of different pad shapes on the component and solder sides. If there are different pad shapes for identical holes, the traveler must call for the use of two separate solder mask films. A note on this must be made for the photo department and the solder mask operators.

c. The customer may have submitted incomplete artwork or may have mixed inner layers from one part number with artwork from another part number. Unless the planner and artwork inspector check and count each piece of film, such errors may go unnoticed at first.

d. Surface mount jobs require special solder mask film. The solder mask cannot be created from a pad master alone. If the job is a surface mount, there must be two pieces of artwork for solder mask. Customers often neglect to submit one or more pieces.

What is the difference between a negative and a positive? This distinction should be clear to everyone handling film. In the printed circuit world:

a. A positive is artwork on which the circuitry portion is dark and the bare laminate portion is clear. The area to be metal is dark (solid).

b. A negative is artwork on which the circuitry portion is clear and the bare laminate portion is dark. The area to be metal is clear.

c. Legend
   (1) Positive legend artwork has dark characters and lines on a clear background. The area on the artwork which will form epoxy characters and lines on the board is dark (solid).
   (2) Negative legend artwork has clear characters and lines on a dark background. The area on the artwork which will form epoxy characters and lines on the board is clear.

d. Solder mask
   (1) On positive solder mask film, the portion which is to be solder mask is dark (solid) and the portion which is to be reliefs around holes is clear.
   (2) On negative solder mask film, the portion which is to be solder mask is clear and the portion which is to be reliefs around holes is dark (solid).

If the area which will be metal is dark, the artwork is considered positive. This is true of power and ground layers of multilayer boards as well. Positive artwork for a ground plane will be dark, with clear areas at the holes. Negative artwork for internal power and ground layers will be mostly clear, with dark spots at the holes.

5. Registration. Registration must be checked for every layer of film supplied. Generally, one layer of film is chosen as the reference and all other pieces

are checked against it. For a double-sided board, the side chosen for a reference is usually the side shown on the blueprint, which may be either the component or the solder (circuit) side. For a multilayer board, the side chosen for a reference is always layer 1. The layer 1 side may be either the component or the solder side. About 70% of the time, layer 1 is the component side; about 30% of the time, it is the solder side. Regardless of which side is shown on the blueprint, layer 1 should always be used to check registration of the other layers, including solder mask and legend artwork.

When checking registration of one layer to another, it is necessary to align one piece of artwork atop the other. It is unreliable to register two negatives or two positives when both are made from silver halide. It is just too difficult to tell the upper and lower pieces of artwork apart. It is best to make either a silver halide reversal or a diazo copy of the reference layer. A negative and a positive can be reliably aligned and perfect registration verified or misregistration measured. Similarly, a diazo copy can easily be distinguished from silver halide. A diazo copy is probably easier to use. It also facilitates inspection for other conditions, such as missing pads, holes hitting traces, and other problems which will be discussed. Any time a reversal or diazo copy is made, it is necessary to check registration between the original and the copy. If registration is not perfect, do not use the copy. The cause of the misregistration must be determined and corrected. Never shoot a copy of the customer's artwork and use it for artwork inspection immediately afterward. Registration must be verified and assured before using this copy.

Once a reversal or diazo copy of the reference layer is obtained, it must be checked against the artwork for the other layers, legend, and solder mask. The pad master should also be inspected for registration if it will be used for any purpose, such as drill programming or generating solder mask or flat mask (the tin-nickel/flat mask process is discussed in Chapter 5). The layer to be inspected is taped to a light table. The reference layer copy is aligned and registered to it. The reference layer should also be held in place with tape for inspection and measurement. When aligning one piece of artwork over another, it is helpful to place your hands together in the center of the film and move them outward to force air from between the pieces of film. This will keep the film from sliding and improve contact for better registration.

In the ideal situation, all layers of artwork align perfectly to the reference artwork, with no measurable misregistration. If the film does not align perfectly, the amount of misalignment must be registered. Before taking a measurement (a 60× microscope is almost an industry standard), the artwork should be aligned along one edge. This allows measurement of the misregistration at the farthest location—the other side of the film. Do not "split the difference," dividing the misregistration evenly, left to right, from the center outward. The goal is to measure worst-case, total misalignment (see Figure 4-2). When the artwork misaligns, there will be a crescent of light where one pad misregisters to the corresponding pad on the other layer. The width of this crescent is the amount of misregistration to be recorded (see Figure 4-3).

Sometimes almost all the pads of one layer align perfectly with the reference artwork but the pads of the B holes, for example, misregister. If only one set of holes misregisters, a note so stating should be made on the incoming inspection report and the amount of misregistration should be specified. Obviously, whenever registration seems to be perfect, an inspection must be made for isolated cases. In fact, even when misregistration is present and measured, the problem of isolated cases remains. It is important to make a distinction between total misregistration, left to right across the circuit, and isolated misregistration. Each condition causes its own problems during manufacturing and must be handled separately.

The inspection should be continued in a like manner for each piece of

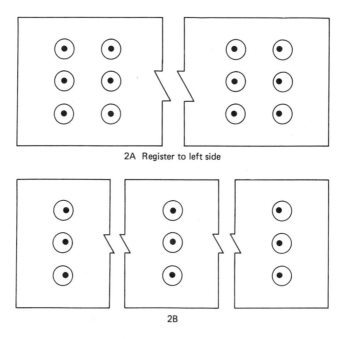

2A Register to left side

2B

Fig. 4-2 Verifying misregistration. (A) Misregistration should become progressively more apparent from left to right. (B) Do not split the difference from the center outward when artwork misregisters and a measurement must be taken.

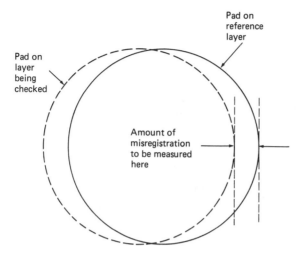

Pad on
reference
layer

Pad on
layer
being
checked

Amount of
misregistration
to be measured
here

Fig. 4-3   Where to measure the shift due to misregistration.

artwork provided by the customer. If isolated misregistration occurs on one layer, inspect for it on the others as well. It is common for only one piece of artwork to misregister and for the others to be fine. It is also common for isolated misregistration to occur on one layer only, or on all of the other layers. The incoming inspector must be thorough and take an adequate amount of time. There are other conditions which can be inspected for at this time or after registration has been determined for all layers. The measured amount of misregistration should be entered on the incoming inspection report in the column titled "Registration." A value should be entered for each layer. If registration is perfect, enter .000 for that layer.

When registering legend artwork, the inspector should also notice whether or not the image will fall on pads or directly across holes. Both of these conditions must be noted on the incoming inspection report. One condition will result in legend ink getting into holes; the other will result in legend ink on pads. Either may contribute to an electrical testing problem when the legend is being applied to the solder side of the board. Special attention must be paid to the surface mounting pads, as legend ink on them should be avoided.

6. Minimum Trace Width. This is important information. Special attention is paid to minimum trace width because this is one of the factors which determines the difficulty of a job. The incoming inspector must look for and record the minimum trace width in the appropriate column on the incoming inspection report. Power and ground layers generally do not contain traces. They do, however, have tie bars connecting selected holes to the power or

ground layer by thermally relieved connections. On internal power and ground layers, the minimum trace width refers to the width of the tie bar. This provides information that the planner needs if it is necessary to increase clearance diameters at these layers. On legend artwork, minimum trace width refers to the width of the lines making up characters. This is important for the planner to know, as character line widths less than .007 inch do not print well by screening methods.

7. Air Gap or Conductor Spacing. This is divided into three separate columns: pad to pad, pad to trace, and trace to trace. When thinking of conductor spacing (air gap), there is a tendency to consider only the spacing between traces. However, many of the problems created by artwork deal with the spacing from pad to trace and from pad to pad. It is the pad-to-trace air gap which determines how difficult solder mask application will be. Each condition must be reviewed carefully by the incoming inspector (see Figure 4-4). One must also be aware of a special type of pad to trace air gap. When a board contains surface mount pads, it is a fairly common design practice to run traces between the surface mount pads. These pads are typ-

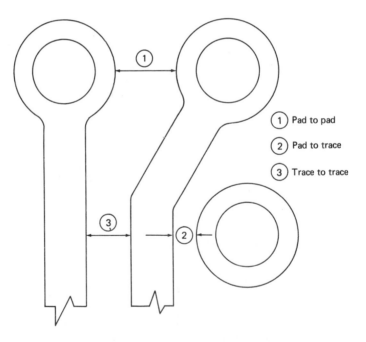

Fig. 4-4 The three types of air gap (conductor spacing) where measurements must be taken: (1) pad to pad, (2) pad to trace, (3) trace to trace.

ically on .050-inch grid centers. Printed circuits designed for surface mounts with a conductor running between these pads present the most challenging solder masking condition possible. It is even worse when the board is being built SMOBC. Yield, quality, and customer–vendor relations would improve if designers avoided this practice and ran an additional layer instead. The additional layer would allow pads only on the outer layers (see Figure 4-5).

8. Annular Ring and Inner Layer Clearance. These are similar in nature and in method of determination. Table 3-2 in the previous chapter is reproduced here. This table (see Figure 4-6), together with the instructions, shows how annular ring and inner layer clearance are determined. Until the incoming inspector has gained a great deal of experience at this task, it is wise to construct such a table for each job. When the incoming inspector is experienced, all that is necessary is to record the measured values and enter them in the chart at the bottom of the first page on the incoming report.

The determination of annular ring and inner layer clearance involves finding the difference in diameter between the drilled hole and the pad or inner layer clearance around that hole. If the difference in diameter is divided in half, the resulting value is the raw annular ring or inner layer clearance. The term "raw" is used because two other factors affect the

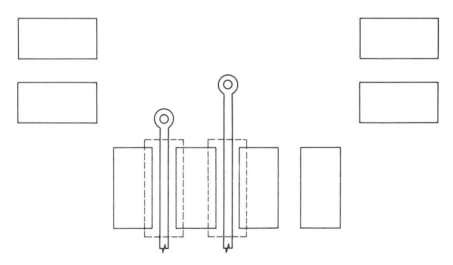

Fig. 4-5 SMD solder masking difficulties. The boxed-in areas (dashed lines) are difficult to solder mask satisfactorily. Mask on pads and exposed traces are common when traces run between SMD pads.

CHARTS FOR CALCULATING ANNULAR RING AND INNER LAYER CLEARANCE

FOR ANNULAR RING:

Hole Code	Pad Size	Drill Size	Mis- Registr.	Shop Tol.	Result	Result = 2	Annular Ring
----	----	-----	--------	----	------	----------------	
----	----	-----	--------	----	------	----------------	
----	----	-----	--------	----	------	----------------	
----	----	-----	--------	----	------	----------------	
----	----	-----	--------	----	------	----------------	

FOR INNER LAYER CLEARANCE:

Hole Size	Clrnc Dia.	Drill Size	Mis – Registr.	Shop Tol.	Result	Result = 2	Inner Clearance
----	-----	-----	--------	----	------	--------------------	
----	-----	-----	--------	----	------	--------------------	
----	-----	-----	--------	----	------	--------------------	
----	-----	-----	--------	----	------	--------------------	
----	-----	-----	--------	----	------	--------------------	

Instructions:
1. Hole Size: for each hole code. write in the nominal hole size.
2. Measure the pad or inner layer clearance for each of the hole sizes. and write this value under Pad Size or Clearance.
3. Write in the diameter of the drill bit which will be used for each hole size.
4.   Write in the amount of misregistration measured. If outer layer pads are being measured. this value will be the front to back misregistration. If inner layer clearances are being measured. or inner layer annular ring. the value entered will be the worst case inner layer misregistration measured.
5. Write in the shop tolerance currently being used. This will most likely be +.002 or +.003 inch.
6. From the Pad or Clearance Diameter. subtract: Drill Size
                                                 : Misregistration
                                                 : Shop Tolerance
    The value obtained is the Result.. This is the difference in diameters of the Pad/Clearance and the drill size. after all registration factors have been subtracted.
7. Result/2 = Annular Ring. or inner layer clearance — after all registration factors have been subtracted. This is the actual value likely to be measured on the final printed circuit board.

Fig. 4-6 Charts for calculating annular ring and inner layer clearance.

actual clearance or annular ring: layer-to-layer misregistration and shop processing tolerance.

Any layer-to-layer misregistration has to be subtracted from the raw annular ring. If the raw annular ring is .010 inch but registration is off .002 inch, the annular ring can be no better than .010 − .002 = .008 inch for the layer where the misregistration occurs. The .008 inch will give a better approximation of what the actual annular ring will be on the finished product than the difference in diameter alone.

The shop processing tolerance cannot be neglected either. Some shops allow +.002/−.002 or +.003/−.003. This means that they expect to hold all location tolerances within this value. All induced misregistration from programming, drilling, photo and imaging operations, artwork changes, thermal processing of laminate, and human error contribute to the shop tolerance. This is a real value and cannot be neglected. If the raw annular ring is .010 inch, this value is reduced to .008 inch after subtracting the layer misregistration from the customer's artwork. Now it is time to subtract the shop tolerance. This reduces the actual annular ring an additional .002 or .003 inch (depending on which value is allowed for shop tolerance). In this author's experience, .002 inch can be held only a portion of the time. A shop tolerance of .003 inch will more accurately reflect what occurs during manufacturing. The closer the shop can come to .002 inch, the better their controls. Thus, for the case in point:

$$\text{Annular ring} = \text{raw annular ring} - \text{misregistration} - \text{shop tolerance}$$
$$= .010 - .002 - .003 = .005 \text{ inch}$$

There is no reason for a shop to produce a circuit with an annular ring of less than .005 inch. If there was no misregistration on the incoming artwork, the annular ring which could be expected would be:

$$\text{Raw annular ring} - \text{shop tolerance} = \text{actual annular ring}$$
$$.010 \text{ inch} - .003 \text{ inch} = .007 \text{ inch}$$

In determining the annular ring, two conditions are sometimes noticed: breakout and wipeout. Breakout occurs when a resulting annular ring is less than specified. This may be due to inadequately sized pads for a given hole or misregistration. Wipeout occurs when the drilled hole is equal to or greater than the pad size. Both conditions must be questioned for disposition on the incoming inspection report. There is a special case of breakout or wipeout, that incoming inspectors should be aware of. Sometimes the drilled hole wipes out or breaks out of the pad and touches a trace. The hole is so much greater than the pad that it creates a hole-to-trace air gap problem. There is no easy solution for this problem. The customer must be advised of it. Possible solutions are to greatly reduce the hole size, greatly reduce the trace size, or reroute the trace(s).

9. Solder Mask Clearance. Determining solder mask clearance is a special case of determining the annular ring. Instead of finding the difference in diameter between the drilled hole and the pad on the artwork, we must find the difference in diameter between the solder mask opening and the pad around the hole (or surface mount pad). To gain experience, the incoming inspector can fill out a chart like the one below. After the incoming inspector has gained experience in making these determinations, he/she need only record the measured values in the chart at the bottom of the first page of the incoming inspection report. The actual calculation is quite simple and can be made mentally.

Hole Size	S/M Clrnce.	Pad Size	Mis-regstr.	Shop Tol.	Result	$\dfrac{Result}{2} = \dfrac{S/M}{Clrnce}$
___	___	___	___	___	___	___
___	___	___	___	___	___	___
___	___	___	___	___	___	___

The surface mount pad is like a hole; an electrical connection will be made, and a component will be soldered there. The solder mask clearance is even more important, since solder mask on pads may cause the board to be non-functional (scrap). The calculation is virtually identical, except that the difference in width and solder mask opening, not diameter, is used to determine the raw annualar ring (see Figure 4-7).

The annular ring, inner layer clearance, and solder mask clearance must be inspected against the inhouse guidelines of the printed circuit shop and the requirements of the customer (including IPC and military specifications). If the annular ring and the clearances are inadequate, this must be questioned in writing on the incoming inspection report. If the solder mask clearance is .015 inch and the pad-to-trace air gap is .008 inch, traces will be exposed by the solder mask; this must be questioned in writing on the incoming inspection report. Dispositions to the problems must be responded to in writing, as well as on the incoming inspection report.

10. Other Items. There are numerous other items to inspect for on incoming artwork. Some of these items can be checked while taking registration measurements, when each layer of artwork is being matched against the reference layer. A diazo reference is a very helpful aid in identifying some of the conditions discussed below.
    a. Inner layer power to ground short. This is inspected for by registering the power and ground artwork over a light table. If any given hole makes an electrical connection to both the power and the ground planes, the two planes will be shorted together. This is usually a design error

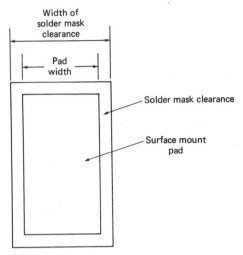

Width of
solder mask
clearance

Pad
width

Solder mask clearance

Surface mount
pad

Fig. 4-7 Solder mask clearance around a surface mount pad.

and must be brought to the customer's attention prior to planning the job. There are times when it is permissible to have the planes connected. If one of the planes has been divided into separate voltage areas, direct shorts are sometimes intended. Planes which have been divided into different voltage areas will have a metal-free channel running through them.

b. Pads on one layer, no pads on another. Sometimes there will be pads on the component side, for example, and none on the solder side. This is usually an error and should be questioned.

c. Pads hitting traces. Sometimes a pad just touches a trace; this condition is called "tangency," (see Figure 4-8). The customer must be consulted for the disposition in such cases.

d. Traces without ends or no pads at the end of the traces. Sometimes this is done to provide a shield. Most often, however, this condition is a design error and must be rectified by the customer.

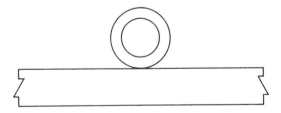

Fig. 4-8 The pad is tangent to the trace.

e. Crossed traces. Sometimes the incoming inspector will notice traces which intersect and then continue on their way. These crossed traces may or may not have been designed this way on purpose. This is the type of condition which should be mentioned to the customer.

f. An air gap problem caused by placing an internal trace too close to the drilled hole. This is a design rule violation. This condition is impossible to detect simply by inspecting the artwork for an internal signal layer. It can be detected only by registering the artwork to one of the following: (1) the drilled panel, (2) the pad master, or (3) the reference layer.

g. No clearance holes or thermally relieved connections to an internal power or ground plane. In this situation, there are more pads on the reference layer than on the internal plane layer. This condition is almost always a design error which must be rectified by the customer.

h. More clearance holes on internal planes than there are holes in the board. This situation is similar to the presence of nonfunctional pads on internal signal layers. It will cause no problem. If there are no other problems requiring the customer's disposition, there is no need to query the customer about extra clearance holes on internal planes. It is important, however, to verify that the layer being inspected is the correct part number and is the same as the rest of the artwork.

i. Clearances in the solder mask artwork where there are no pads on the board. If the extra clearances expose traces or metal planes, the customer must be consulted for disposition. If the extra clearances simply expose laminate, there is no cause for concern.

j. The layer sequence on the artwork is incorrect and/or does not match the blueprint's layer diagram. The customer must be consulted.

k. The blueprint must identify the side showing as the component or solder side, or must use other terms to present this information. Multilayer blueprints must also contain information to show which side is layer 1. The planner should already have verified this information.

l. Hole counts must be correct. If there are only about 20 holes of a given hole coding, the planner or incoming inspector should have counted them on the blueprint to verify that the count on the hole chart is correct. Inaccuracies should be questioned, as this is a possible design error.

m. Dimensions, especially for dimensioned holes, should be measured to verify that the artwork matches the blueprint. Any instrument from a steel ruler to a digitizer will be adequate for most applications.

n. The incoming inspector must verify that the correct drill size has been called out for each hole size. The drill bit assigned by the planner must also reflect whether the hole is to be plated or nonplated.

o. The quality of the circuitry must not be forgotten. Short circuits, broken traces, excessive spots or pinholes, fuzzy circuit definition, light edges to circuitry, raggedness, and other such conditions must be evaluated for inhouse acceptability.

p. The circuitry shown on the blueprint must be identical to that on the artwork for the side shown. Any discrepancies must be questioned.

q. The hole pattern should be symmetrical. There is a 50-50 chance of building a board with a symmetrical hole pattern backward. Special notes must be written on the blueprint and the traveler to warn of the symmetrical hole pattern. The artwork for symmetrical hole patterns will register to the drilled panel in more than one way; however, only one of these is correct. The artwork may not be symmetrical; this makes no difference. It is the symmetrical hole pattern which is cause for concern.

During the artwork inspection process, when items such as those discussed above are noticed, they should be identified with arrows. If there are several discrepancies, the arrows can be numbered. Not all conditions which pose a problem or a potential problem are spelled out on the incoming inspection report. It is not the intent of the report to provide many squares to be checked off. The report lists measurements which must be performed, as well as some of the most commonly found problems—but not all of them. The incoming inspector must know how to inspect artwork. A simple checklist is not sufficient.

OTHER INCOMING INSPECTION ITEMS

The incoming inspector is responsible for double-checking everything on the traveler and blueprint. This includes all part numbers; quantities; material callouts; quality requirements; multilayer construction, solder mask, and legend requirements; panel serialization (mil-spec jobs); instructions to photo, programming, and other departments; and all of the manufacturing operations listed on the traveler. The incoming inspector must be as familiar with the information discussed in the next chapter as the planner should be with the information covered in this chapter.

Chapter 5

Planning: Multilayer And Double Sided Printed Circuits

THE ROLE OF THE PLANNER

Basic duties of the planner include the following:

1. Review all incoming documentation from the customer.
2. Understand the general requirements of each customer, as made known by the procurement specification and the working relationship developed with that customer.
3. Review the documentation on each job for requirements and manufacturability.
4. Transmit all manufacturing, quality assurance, and documentation requirements of the job to the rest of the shop's personnel.
5. Identify, prevent, and resolve problems associated with each job.
6. Maintain records on each part number being run inhouse.
7. Maintain records of customer contacts relating to any given job.
8. Know the IPC specifications, military specifications and customer specifications.
9. Be familiar with all manufacturing and tooling operations.
10. Be familar with the various laminates available for printed circuit manufacturing.
11. Understand inhouse documentation for production control, remakes, and returns.
12. Be able to prevent the repetition of problems on later releases of a given part number and on reorders of previously run part numbers.
13. Be able to keep track of laminate and special materials.
14. Deal with outside services, as needed, to supplement the shop's capabilities.
15. Know how to lay out a panel.

The Job Traveler

This is the basic document used by the planner. It is like an artist's canvas. A well-written traveler will communicate all of the requirements necessary to meet the customer's specifications and expectations. Any traveler must contain the following basic information:

1. Documentation information about the customer and part number
   a. Customer
   b. Part and revision numbers
   c. Purchase order number
   d. Due date
   e. Quantity due and start quantity

2. Panel information
   a. Panel size
   b. Panel layout
   c. Type of laminate
   d. Number up per panel

3. A sequential listing of each operation, including first-article inspections not listed below
   a. Planning
   b. Incoming/artwork inspection
   c. Tooling
      (1) Drill programming and first-article preparation
      (2) Router programming and first article
      (3) Test fixture programming and drilling
      (4) Photo preparation of working film
   d. All manufacturing operations, including their requirements
      (1) Laminate shearing: double- and single-sided boards
         (a) Inner layer operations
            • Shear
            • Clean (chemical or mechanical)
            • Dry film coat
            • Punch tooling holes (or drill)
            • Inspect/touch up inner layer working film
            • Expose inner layers
            • Develop inner layers
            • Touch up
            • Etch
            • Inspect

- Black/brown oxide
- Bake
- Laminate/press
- Post lamination bake

(2) Drilling
(3) Electroless copper
(4) Imaging
(5) Touch up
(6) Plating
(7) Resist strip
(8) Etch
(9) Contact finger plate
(10) Reflow/fuse
(11) Inspect
(12) Apply solder mask and legend
(13) Fabrication (board outline routing)
(14) Final cleaning
(15) Final inspection
  - Visual
  - Electrical testing
  - Microsectioning/Group A mil-spec
  - Source inspection
  - Hardware installation or minor assembly
(16) Packaging
(17) Shipping
(18) Track overage and production control functions

4. A place for operator signature, date, and panels counts at each operation, including a location for reporting scrap as it occurs at each operation

5. A location for writing explanatory notes

DISCUSSION OF REQUIREMENTS

1. Documentation Requirements. This category includes the items listed above, as well as those listed in the previous chapter on artwork and incoming inspection. Prior to writing up the job traveler, the planner must review the customer's procurement specification or spec summary sheet. He/she must also review the blueprint and artwork for basic information and tolerances. This is part of reviewing manufacturability. To facilitate the documentation review, the planner should also initiate the artwork/incoming inspection form. The planner must hold the sales department accountable for and verify

that all the information received for a job is accurate and complete, including the following:

a. Knowing the revision level.

b. Submitting forms with complete part numbers. If the actual part number is, for instance, 300-233309-002, the planner must not allow sales order forms with only 233309 listed. In addition, the part and revision numbers on the sales order form, blueprint, and artwork must be identical. If there are discrepancies, the planner must resolve them. When there is no apparent reason for the difference, the planner must question the salesperson. The traveler should have the part number and revision level divided into three categories; as on the incoming inspection report.

c. Submitting new revisions as new revisions, not as repeat jobs. New revisions must have tooling charges included. Even if the reason for the revision is minor and prior tooling will be usable, tooling may have to be charged. After all, the planner, the incoming inspector, and one or more supervisors have spent valuable time trying to identify the changes.

d. Running a repeat as a repeat, rather than as a new job. If a repeat is run as a new job, there will be two sets of artwork, drill/route tapes, and test fixtures for the same part number. This is costly and time-consuming, and can lead to confusion downstream.

e. Remembering to charge for electrical testing and remembering to place the requirement on the sales order form. The planner must know the customer's specifications. It is very common for companies to state that all boards, or at least all multilayer boards, require electrical testing—whether this is on the purchase order or not. Sometimes a salesperson is not even aware that testing is required. Even if the board manufacturer does not charge for electrical testing, this must be listed as a requirement on the sales order form.

f. The name of the buyer and the buyer's telephone number must be on the sales order form. Some companies have several divisions; others have several buyers for printed circuit boards within the same division. When the planner has to contact the customer to resolve a problem, he/she should be able to refer to the sales order form for a name and telephone number.

g. Billing and shipping addresses should be clearly identified.

h. Each sales order form must have quantity and due dates clearly identified. If there are multiple releases on one purchase order, the dates and quantities must be spelled out.

i. The blueprint should identify the side being shown and should specify which side is layer 1 for multilayer boards. Fabrication tolerances, hole tolerances, foil thickness, laminate type, dielectric spacing requirements, and other items should be reviewed. However, the planner can wait for the full artwork/incoming inspection report before going into detail on

these items. The established shop guidelines for determining manufacturability should be kept in mind at all times. The planner is the person
who should initiate the incoming inspection report for each job. While
reviewing the artwork and blueprint, the planner can make a note of
problems to be resolved when the full incoming/artwork inspection is
completed.

j. The quote should be reviewed. The planner should take a few extra minutes to make sure that the pricing information is correct. A quote summary form should have been submitted with the sales order form. The
quote summary form shows what the quote was based on. Often the
salesperson will take information about the part number over the phone,
without having a blueprint or artwork to look at. Sometimes the blueprint
is available but lacks full information, such as the fact that the part is a
multilayer or contains gold contact fingers. A quote can be given over
the phone or even submitted in writing. Several weeks may elapse between the time when the quote is given and the time the purchase order
is issued. The salesperson may never even look at the complete document
package when the purchase order is issued. It is fairly common for significant facts about a job to slip by. As part of the documentation review,
the planner must review the quote. If the price seems out of line for the
job, this matter should soon be pursued.

2. Panel Layout Considerations
    a. Generally, one should lay out as many circuits on a panel as possible for
medium-sized to large jobs. This speeds processing through the shop and
makes better use of materials, equipment, and people.
    b. For small jobs, especially prototype and quick turnaround/premium jobs,
it may be better to run with fewer up on a small panel size in order to
reduce the amount of time in the photo department.
    c. Panel layout for small boards (around 2 × 2 or less) where the artwork
comes in one up:
        (1) For quick turn around and prototype (small quantity), run with fewer
up on a smaller panel. This may result in a small panel or a panel
with some unused area. However, the resulting savings in photo
time and cost will offset the poor panel area utilization.
        (2) For larger runs with more lead time and a greater likelihood of
being reordered, use the larger panel size. The workload in the photo
department will be a key factor in making this determination. Check
with the manufacturing manager.
    d. Small boards whose artwork has had step and repeat can be run on larger
panel sizes, since step and repeat will be of little trouble here. If incoming artwork is received with multiple patterns, this multipattern copy can

be stepped even further with no difficulty to make optimum use of the panel sizes available.

e. In general, multilayer boards should follow the same circuits per panel layout rules as double-sided boards, regardless of the number of layers involved. In several cases, a planner may consider running fewer multilayer circuits up on a panel than would be run for a double-sided part number. These cases involve (1) available tooling plates, (2) restrictions on inner layer clearance for power and ground layers, and (3) restrictions on solder mask clearance. Multilayer panel sizes are often limited to a handful of standard sizes. The available panel sizes are dictated by the tooling plates available inhouse. Obviously, the planner must consult his/her list of available multilayer panel sizes. There are times when inner layer clearance is less than what the shop would prefer—for instance, a hole-to-ground plane clearance of .010 inch. It is advisable to limit misregistration due to dimensional changes in artwork and laminate. Thus, a multilayer with tight clearances may be run with a reduced number up, per panel, and on a smaller panel size. Tight solder mask clearances can be handled with greater ease when the panel size is small and there are fewer circuits per panel. This is discussed in Chapter 3.

f. Repeat jobs for medium-sized to large quantities should be considered for relayout if the initial runs have been for small quantities on small panel sizes. This allows the shop to make better use of its materials and facilities. The manufacturing manager should be consulted before the planner issues instructions to lay out a job again. This process will impact programming, drilling, fabrication, F/A inspection, and photo work. Verify that these areas can handle the load before giving this instruction.

Note: Military jobs to MIL-P-55110D: Mil-spec jobs present a unique consideration. All panels on military multilayer jobs must be microsectioned a minimum of two times per panel. There must be a 100% hole quality examination and a 100% thermal stress examination of the plated holes. The number of microsections is determined by the number of panels, not the number of boards. If 100 boards can be built on one panel, only two microsections need be performed. If 100 boards are built on 10 panels, 20 microsections must be performed. If 100 boards are built on 50 panels, 100 microsections must be performed and evaluated. Microsectioning is costly and time-consuming. Quantities for minimum prototype jobs are of little concern. However, short production runs, and larger ones, should definitely be run as many boards to a panel as possible.

g. The optimum panel size and the number up are the primary concerns for panel layout. However, after these values are determined, there is another consideration. Even numbers can be flip-flopped; odd numbers cannot. The following must be considered:

(1) If both sides are solder masked, flip-flop will reduce screening setups and oven times by 50%.

(2) If a legend is required on both sides, flip-flop will reduce screening setups and oven times by 50%.

(3) If a multilayer is to be built, flip-flop will reduce step and repeat time, in photo, film cost, and labor, by about 50%.

(4) If the job is to be screened for pattern plating, flip-flop will reduce screening setups and photo time by approximately 50%.

(5) Jobs with gold plated contacts are a special case. The contact fingers must be aligned along the outer edges of the panel. If a panel must be sheared into two or more smaller panels, the photo department must be informed to set up solder mask and legend artwork to accommodate the reduced panel size, not the unsheared size (see Figure 5-1).

(6) Military jobs can be run with even or odd numbers up on a panel, since the patterns should not be flip-flopped under any circumstances. This will be discussed below.

h. Single-sided boards must be run straight up, since there is copper on only one side of the panel. Boards classified as single-sided with plated-through holes (which are actually double-sided boards) must also be run straight up. This allows the platers to rack and plate these panels back to back. Racking back to back will reduce the tendency toward funnel plating, whereby the hole reduces very rapidly on the side of the board with pads alone.

i. Multilayers, like double-sided boards, should be run flip-flop, with the following limitations:

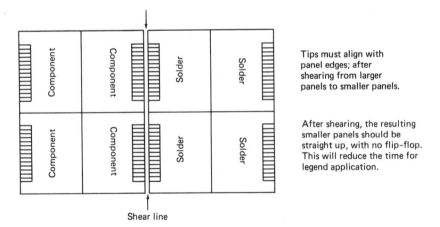

Fig. 5.1 Panel layout considerations for boards with gold contact fingers.

(1) If the best panel layout includes running an odd number of circuits, flip-flop will not be possible (see Figure 5-2).

(2) If the multilayer has an odd number of layers, the circuits can still be flip-flopped. However, many circuits with an odd number of layers have dielectric spacing requirements between those layers. Restrictions on dielectric spacing may make it impossible to flip-flop the circuits (see Figure 5-3).

(3) If the layer spacing is uneven do not flip-flop. (This case is like that above.) Restrictions on dielectric spacing, such as unequal layer spacings, make it impossible to flip-flop.

(4) Uneven internal copper foils make it impossible to flip-flop circuit patterns. For example, an eight-layer board with 2-oz foil only on layer 2 cannot be flip-flopped. If it were, there would be a mixture of 1- and 2-oz foils for layers 2 and 7 (see Figure 5-4).

(5) For mil-spec jobs and other jobs requiring coupons for determining

Comp	Comp	Solder	Solder
Solder	Solder	Comp	Comp

Even number of circuits can be set up for flip/flop.

(a)

Comp	Comp	Comp
Comp	Comp	Comp
Comp	Comp	Comp

If the best panel lay-out results in an odd number of circuits per panel, the images cannot be flip/flopped. The only drawback is a little more time in photo and screening.

(b)

Fig. 5.2   (A) An even number of circuits can be set-up for flip-flop. (B) If the best panel layout results in an odd number of circuits per panel, the images cannot be flip-flopped. The only drawback will be a little added time in the photo and screen printing departments.

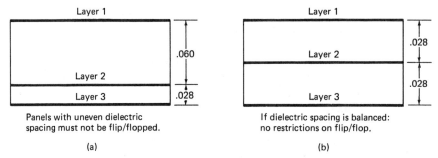

Fig. 5.3 (A) Panels with uneven dielectric spacing must not be flip-flopped. (B) Dielectric spacing is balanced; no restrictions on flip-flop.

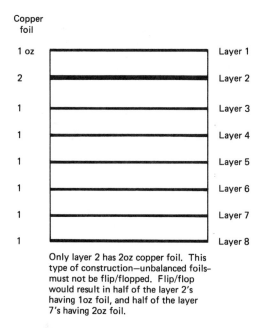

Fig. 5.4 Unbalanced foil. Only layer 2 has 2-oz foil. This type of construction—unbalanced foil—must not be flip-flopped. A flip-flop layout would result in half of layer 2's having 1-oz foil and half of layer 7's having 2-oz foil.

layer-to-layer registration, layer-to-layer dielectric spacing, and inner layer foil thickness, do not flip-flop. Flip-flop is technically permitted. No design or manufacturing rules will be broken. However, it will be very difficult for the laboratory to distinguish, for example, layer 1 from layer 10 on a ten-layer board. It can also create difficulties in identifying dielectric spacing and specific inner layer copper with the layer number. Laboratory people will find it easier to work if all mil-spec jobs are run straight up instead of flip-flopped.

j. Other layout considerations

   (1) Allow a 1-inch border inside the panel edges for the plating border and for multilayer tooling holes.

   (2) Spacing between circuits should be as follows:

   • Between multilayer images, allow .500-inch spacing. This provides channels for air removal and resin flow. It is possible to run with .200-inch spacing in most cases, especially when the dot/bubble pattern is used for inner layer borders. However, there are times when air entrapment (resembling delamination) occurs with .200-inch spacing, which would not have happened with .500-inch spacing.

   • Between double- and single-sided images allow .200-inch spacing.

   • Some panels will be sheared into two or more smaller ones, such as panels being sheared for testing, tip plating, or solder masking. In these cases, allow .500-inch spacing between the circuits where the shearing cut will be made. Allow .200-inch spacing between the other circuits on the panel.

   (3) Important note: Extra screen holes are required when a panel is to be sheared into smaller panels prior to solder mask and legend application. The resulting smaller panels must be programmed and drilled to have the required five screening holes in the corners after shearing. The planner must instruct the programming department to add these holes on all jobs which will be sheared (see Figure 5-5).

   (4) Test coupons require certain layout considerations. MIL-STD-275 coupons are about .500 inch across and several inches long. The actual length depends upon the number of layers involved. However, the entire .500-inch width should be placed inside the 1-inch plating border. The coupon is just as important as the boards on the panel. The microsectional quality of the coupon will be used to judge the acceptability of all the boards built on a panel (see Figure 5-6).

   NOTE: Whenever a job is being run as MIL-P-55110, or whenever the coupons will be used to judge acceptability, the panels must be serialized prior to fabrication. The coupons must be saved and given to the laboratory for microsectional or Group A evaluation.

k. Jobs which fall into one of the following categories should be planned and programmed for shearing into smaller panels at the appropriate time.

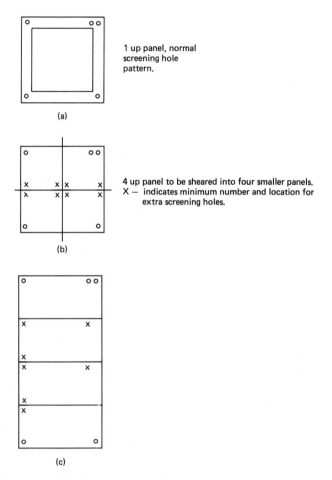

Fig. 5.5 Extra screening holes must be drilled when a panel is set up for shearing. (A) One up, normal screening hole pattern. (B) four up; this panel will be sheared into four smaller subpanels. X indicates the minimum number and locations of extra screening holes that are required.

(1) Solder masking of large panels
  • SMOBC jobs where the artwork report states that the solder mask clearance is less than .010 inch, after all registration factors have been subtracted. These panels should be sheared in half for solder masking.
  • Non-SMOBC jobs where the solder mask clearance is less than

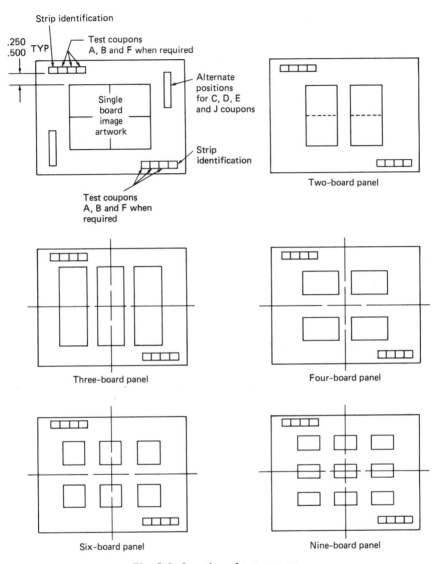

Fig. 5.6   Location of test coupons.

.008 inch after all registration factors have been subtracted and where the customer will not allow mask on pads. These panels should be sheared in half for solder masking.

- Heavy copper foil (2 oz or more) is likely to produce skips. If the solder mask clearance is .010 inch or greater, these types of panels should still be set up for shearing in half because the stencil will need to be heavily flooded.

NOTE: We look for .010-inch solder mask clearance on SMOBC jobs and only .008-inch clearance for other jobs. This reduces the possibility of getting solder mask on the pads before solder leveling.

(2) Contact fingers away from the panel edges require that the panel be sheared.

Typical Planning Procedure

The procedure described below is typical of the work of a planner when planning a job. Some of the values listed for inhouse guidelines are arbitrary. Each shop needs to decide what it can reliably handle. If this has not been determined, the values given here can be used until it becomes apparent that other values should be substituted.

1. Incoming Procedure. When a job order comes in from the sales department, the planner should review the customer order form. Typical information on the customer order form includes:
   a. Customer name
   b. Part number
   c. Revision level
   d. Job number
   e. Delivery date
   f. Number of releases with quantities and dates
   g. Status: new, new revision, repeat, change
   h. Type of processing: SMOBC, tin-nickel/flat mask, premium, multi-layer, mil-spec.
   i. Buyer's name and phone number
   j. Sales person's name
   k. Shipping instructions
   l. Cost, tooling, testing, and premium

2. New and New Revision Jobs. If the job is new or a new revision, the files should be checked immediately to verify this fact. Discrepancies should be immediately brought to the attention of the sales department.
   a. New jobs and new revisions must have tooling charges.
   b. The customer order form must state the bare board testing charges, if

testing is required, and who is to perform the test. Testing should always be performed inhouse. If another source is listed for testing, this matter should be pursued with the sales department.

c. If the job is for a new customer, the manufacturing specification of that customer must be included with the sales order form. Special requirements, such as microsections, solder samples, shipping mode, and any kind of certification, should be noted by the planner and referenced on the traveler.

(1) Any note on the customer order form which states a manufacturing requirement for the job (such as hardware, trace width, or minimum dielectric) should be redlined onto the blueprint. The purchase order (sales order form) takes precedence over the blueprint.

d. All copies of the blueprint should be stamped with the date received.

e. If the job is a new revision, it must be planned as if it were a new job. From this point on, there are two possibilities, depending on shop policy.

(1) The use of old documentation should be avoided. The old revision blueprints, artwork, and N/C processing tapes should be picked up and segregated to prevent their use.

(2) Alternatively, the film package, blueprints, and drilled First Article for the previous revision should be submitted to the incoming inspection department, together with the documentation on the new revision. The incoming inspector should determine if any of the previous tooling (artwork, drill and route tapes, and test fixture) can be used on the new revision.

3. Incoming Artwork Inspection Report. This report should be initiated by the planning engineer.

a. The customer name, part number, and revision level for all documents should be listed, together with the status of the job (new, new revision, etc.). Discrepancies should be questioned. If the part number or revision level is not identical, but is correct, a confirming note should be made so that other departments will not question the discrepancies.

b. The planning engineer should check the artwork to make sure that the part number is correct and that all items are present. This inspection also affords the planner the opportunity to note anything unusual about the film prior to beginning the actual planning operation.

(1) During planning, any items which need further analysis for disposition should be questioned in the artwork inspection report at this time. After the artwork inspection has been completed by the incoming inspector, it is returned to the planner. The planner reviews

all information and measurements on the artwork inspection report. Any items which do not meet established inhouse guidelines or other specifications should be dispositioned by the planner. All dispositions and instructions should be written on the incoming/artwork inspection report.

(2) Jobs with one or more outer layer surfaces that are primarily ground plane should be questioned for possible tin-nickel/flat mask processing. This note should be made on the artwork inspection form and dispositioned after completion of incoming/artwork inspection.

4. The traveler. This should be filled out with all of the guidelines and requirements of the purchase order, blueprint, customer specification, IPC and military specifications, and inhouse criteria in mind.

   a. Inhouse criteria are established to reflect IPC and military standards. All customer documentation and requirements should be judged against inhouse criteria for acceptability and manufacturability. If the customer's requirements and artwork do not meet inhouse criteria, the customer's requirements should be questioned for later disposition according to guidelines established in this procedure.

      (1) Specification summary sheets (specification summaries) are a condensed version of the customer's specification. These should be consulted for each customer prior to planning the job.

5. Inhouse criteria.
   a. Panel size: maximum, 18 × 24 inches; minimum, 9 × 12 inches
   b. Panel layout: refer to item 2 above, "Panel Layout Considerations"
   c. Copper foil: minimum, $\frac{1}{2}$ oz; maximum, 2 oz
   d. Silk screen capability: line width, .008 inch minimum; air gap, .008 inch minimum; annular ring, .002 inch minimum
   e. Dry film capability: line width, .006 inch minimum; air gap, .005 inch minimum; annular ring, .002 inch minimum
   f. Solder Mask: .010 inch minimum clearance at round pads and surface mount pads

   NOTE: Solder mask for surface mount jobs
   Some designers and printed circuit shops set up artwork for surface mount sides so that no solder mask will be applied between adjacent pads. Instead of a relief around each pad, the entire row of pads has one large relief around it (see Figure 5-7). This reduces the total number of places where mask can get on pads. When a trace runs between adjacent pads, the artwork has a solid line to cover that trace. When screening operators are registering the stencil, they need only pay close attention to those areas which are most difficult. If incoming artwork has not been set up this way, the planner should ask the customer

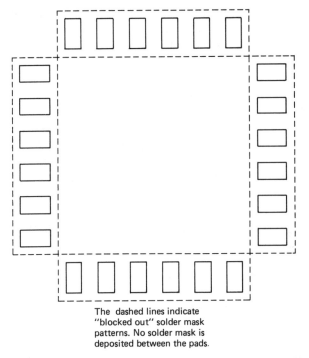

The dashed lines indicate
"blocked out" solder mask
patterns. No solder mask is
deposited between the pads.

Fig. 5.7   Solder mask patterns for most surface mount jobs. The dashed lines indicate blocked-out solder mask patterns. No solder mask is deposited between the pads.

for permission and then instruct the photo department to "block out" the surface mount pad rows in this manner.

g.   Legend
   (1)   Legend characters must have line widths of .007 inch or greater.
   (2)   No legend ink should be allowed in holes or on surface mount pads.
h.   Air gaps (conductor spacing)
   (1)   Trace-to-pad and trace-to-surface mount pad air gaps tighter than .013 inch should be questioned for possible mask on pads or exposed traces. Jobs with pad-to-trace air gaps under .013 inch and trace widths of .010 inch or greater should be reduced at least .002 inch if they are tin-nickel or being screened for plating.
   (2)   Screen printing for pattern plating should not be performed on jobs with the following conditions:
      • Solder mask is required

                    and

      • The pad-to-trace air gap is less than .015 inch after traces have been reduced to compensate for growth during plating.

- The air gap at any point is less than .008 inch after photo reduction of the trace width.

i. Panel thickness and tolerance: .062±.005 inch; .093±.007 inch; .125±.009 inch. If contact fingers are present, thickness should be measured at the contact fingers, unless otherwise specified on the drawing or in the specification.

NOTE 1: All panels that are .093 inch or thicker should be run with tin-nickel beneath the solder unless they are SMOBC. (This is optional.)

NOTE 2: All double-sided panels that are .093 inch or thicker should be set up for smear removal prior to electroless copper application. (This is not optional.)

The procedures described in notes 1 and 2 ensure better hole wall quality and metal adhesion.

NOTE 3: All panels which are .093 inch or thicker should be set up for .0015-inch minimum copper plating. The plated-through holes holes should be drilled at least .006 inch over the middle of the spread for ±.003-inch tolerance and the platers instructed to reduce holes by .004 inch. This compensates for the hourglass plating effects of thick panels.

j. Dielectric and tolerances: .005-inch minimum with ±.002 inch for multilayer laminate min. inch tolerance; >.005 inch with ±.003-inch tolerance. Layer construction should be per inhouse guidelines. Any other callout on the blueprint or in a specification shall be questioned for disposition.

J. Multilayer
   1. Minimum laminate thickness   .005   ±0015 inch
   2. Minimum dielectric spacing   +003
      (Prepreg build-up)   .004   −000 inch

k. Number of layers: 12 maximum.

l. Fabrication and drilling tolerances
   (1) Routed panel dimensions: ±.005 inch minimum
   (2) Routed slots: ±.005 inch minimum
   (3) Routed but plated internal slots: ±.005 inch minimum
   (4) Hole-to-hole dimensions: ±.003 inch minimum
   (5) Hole/datum to panel edge dimensions: ±.005 inch
   (6) Nonplated holes: +.002/−.001 inch minimum diameter
   (7) Plated holes: .018 inch minimum with ±.002 inch minimum; .019 to .070 inch ±.003 inch minimum; >.071 ±.004 inch minimum
   (8) Second drilling to be performed for all nonplated holes, except for

special situations; this should be established in agreement with the N/C supevisor and the manufacturing manager.

NOTE: If a given printed circuit manager has little N/C drilling or routing time available for second drilling, it may be advisable to plug as many holes as possible rather than second drill.

m. Warpage: 1% = .010 inch/inch
n. Etching tolerances: 1 oz. foil: trace width ±.002 inch; 2 oz. foil: trace width ±.004 inch minimum tolerance
o. Electrical testing is limited to holes on a .100-inch grid unless the shop has demonstrated that it can handle holes on .050-inch grid centers. The tightest point spacing which can be handled by most testers is rows of .100-inch spaced holes, where the rows are separated by .050 inch and offset from adjacent rows by .050 inch. There is no problem in testing boards where the vias are off .100-inch grid, since the vias are not drilled into the test fixture (see Figure 5-8).
  (1) Surface mount jobs can be tested only at the component holes. Surface mount pads or feed-through (via) hole testing should be done only when the shop has demonstrated its ability to handle this requirement. When SMD pads are to be tested, a dedicated fixture may prove most reliable.
  (2) Electrical testing shall be performed inhouse. Only the first traveler of a multiple release need be set up for drilling the test fixture.
    • Some shops find it more convenient and reliable to assure that the test fixture will be drilled correctly and on time by issuing a traveler just for building the test fixture.

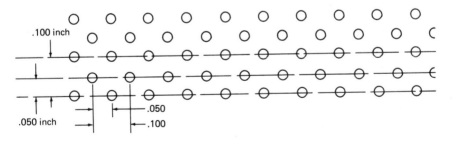

Fig. 5.8 This is the tightest hole pattern that can be tested with most test equipment. (1) The holes in each row are separated by .100 inch. (2) The rows of holes are separated by .050 inch and offset (staggered) by .050 inch.

- Panels which contain a large number of circuits up can be tested in panel form or as quarter panels. If this is desired, instructions to this effect must be given to the programming department. Boards with contact fingers cannot be tested in this manner, as the plating bar to the contact fingers forms a short circuit.

6. The traveler. This should be filled out completely, with any special notes to individual departments needed to enhance understanding.
   a. Holes to be plugged should be identified with red ink on all copies of the drawing. A note should be made to platers at the copper electroplating line on the traveler.
   b. Any redlines to a drawing must be copied on all copies of that drawing. This may entail tracking down jobs already on the floor.
   c. Multilayer steps can be listed separately on the back of the traveler. The back of the traveler thus becomes a separate traveler for processing inner layers through lamination and postbake.
      1. The construction to be used should be planned on the construction diagram. No changes are to be made without notifying the planning department.
   d. Jobs requiring microsections to be shipped with the order shall have appropriate notes written for the premask and shipping departments. A standard inhouse coupon will have to be run. A note must be written for the programmer to include the inhouse test coupon, if this is not already established as normal procedure.

7. Blueprints. The following minimum information must be shown:
   a. Fabrication tolerances
   b. Hole tolerances
   c. The side shown labeled as the component or solder side
   d. Layer sequence for multilayers
   e. Panel thickness
   f. Type of laminate
      If any of the above information is missing, the planning engineer should add it by redlining, signing, and dating.
      All hole charts should have hole codes listed alphabetically. If the customer has failed to do so, the planning engineer should make this assignment. It is easier for platers and other shop personnel to refer to the B holes, for example, than to the holes with the asterisk designation.

8. Discrepancies and Dispositions.
   a. All items questioned by the planning and artwork inspection departments should be dispositioned inhouse, if possible. Both planning and artwork inspection should be completed before attempting to make dispositions.

b. The quality assurance manager or engineering manager should be consulted prior to calling the customer. All customer contacts should be recorded on the artwork inspection report or the blueprint, together with the name of the customer contact authorizing the disposition. Whenever the planner receives a waiver from a customer, the planner should always ask that person for a waiver or deviation approval number. Otherwise, the parts may be rejected at the customer's incoming inspection. Never assume that the customer contact will follow through and inform his/her incoming department.

(1) It is the goal of the planner to limit or eliminate the need to call the customer.

(2) When a buy-off or disposition has been made by a customer, the waiver approval number and the reason for the waiver should be noted on the preshipping inspection form. A copy of this form should be shipped with the job.

(3) The specification summary sheet should be used to document items routinely bought off by customers. This sheet should be consulted for blanket buy-offs already established with a given customer. Whenever possible, these shall be used instead of calling the customer.

(4) The planner should explain conditions requiring dispositions by the customer so that the customer understands fully what is required and what the result will be. Also, the buyer or other contact must understand that delays in response will push out due dates day for day.

9. Special Materials. Special materials, such as hardware, unusual legend or solder mask color/type, heat sinks, test fixtures, and so on should be ordered as the job is released to production.

10. Documentation Package and Traveler. The production control department receives the entire documentation package and traveler for release to the floor only after all planning, artwork inspection, and dispositions have been completed. The only exception is made for premium jobs which may be released for programming at the end of a day, as long as the planning department follows up on the film inspection the next day.

11. Drilling File Card. The planner should fill out a drilling file card for the panel layout, coupon location, and zero dimensions.

a. Nonplated holes should be drilled out near the top the tolerance.

b. Plated-through holes should be drilled out .005 to .006 inch over the middle of the hole spread. If a drill bit is not available, the next larger size should be chosen.

(1) Unusual plating conditions, such as a hole wall copper thickness of .002 inch or greater, require that a larger drill size be used. Note that all panels .093 inch or thicker require a .002-inch plating thickness.

12. Artwork inspection reports. These are to be referenced by personnel performing various operations and must accompany all new and repeat jobs.
   a. When a new job or new revision is released, two copies of the artwork inspection report should be in the traveler. One remains in the traveler at all times; the other remains with the film envelope.
   b. Repeat jobs should have one copy of the artwork inspection report in the traveler when it is released.

13. Job Folder. After a job has been planned and all copies of travelers for later releases have been completed, the planner must make a job folder. Inside the cover of the job folder is information on each run. This information (see Figure 5-9) must be kept current.
   a. When the folder has been made, the planner should sign off on the bottom line of the customer order form for "traveler" and "folder."
   b. The entire documentation package for the part number should now be brought together and taken to the artwork inspection department.
   c. Note that even a remake traveler should be listed in the cover of the folder. The inside cover of the folder shall contain a complete history of each part number.
   d. A requisition should be filled out for special materials, such as hardware or outside testing fixtures (if required), at the time the job is planned and submitted to the purchasing department when the job is released.

| | | | | | | | | | Overage Information | | | |
Job No.	Qty. Due	Unit Cost	Due Date	Qty. Ship.	Bal. Due	P.O. No.		Job No.	Qty. In	Qty. Out	Date	Bal.

Fig. 5.9   File folder information.

14. Special Considerations.
    a. SMOBC
        (1) The solder mask artwork must be set up for a minimum of .020 inch greater diameter than the component and solder side pads. There are no exceptions. The customer must buy off exposed traces if this will result from a pad-to-trace air gap of less than .013 inch.
        (2) SMOBC critical step sequence
            • Strip solder after etching.
            • Pre–solder mask inspection.
            • Solder mask panels: Solder mask should cover the tops of contact fingers, if present. This note must be placed on the artwork inspection form and traveler for all SMOBC jobs.
            • Tip plate contact fingers, if present.
            • Solder dip panels: Hot air level or hot oil level.
            • Route panels to the inspection department.
    b. Surface mount jobs
        (1) A dedicated test fixture should be procured. This will entail substantial cost, which must be reflected in allowance for test fixture, on the sales release.
        (2) If there are traces running between the surface mount pads, the pad-to-trace air gap is likely to be less than .010 inch. Jobs like these need special consideration. If possible, and if permission can be obtained, neck the trace down to .005 inch where it passes between the traces. It may be necessary to set up the panel for shearing into quarters for solder masking.

The Place of the Incoming/Artwork Inspection Report

If the traveler is the planner's most important document or tool, the incoming inspection report is second. This report should contain a list of the most important features of the customer's artwork. It should also contain a list of all problems and potential problems that would be caused by using the artwork or blueprints as supplied. In almost all cases, something must be added to a blueprint or changed on the customer's artwork to make it usable. A full summary of all these items should be given on the incoming inspection report.

The incoming inspection report should contain all of the following instructions to the photo department on how to set up the artwork:

1. Photographic spreads and chokes required.
2. Inner layer clearances and reliefs needed for nonplated holes, slots, cutouts, and internal planes at board edges.
3. Traces which have to be rerouted.

4. Changes needed to accommodate surface mounting, SMOBC, or tin-nickel/flat mask.
5. Areas of special importance, such as traces running between surface mount pads that must be covered by solder mask. Also included are instructions to remove solder mask areas between surface mount pads.
6. Changes required in drill size to accommodate the annular ring or some other consideration.
7. Buy-offs from the customer for solder mask on pads, exposed traces, or annular ring violations.
8. Instructions on touching up the customer's artwork.
9. Special notes on markings and how they are to be applied.
10. Discrepancies on the blueprint and the disposition of those discrepancies.

In addition to above items, any question on the part number that requires queries to the planner or customer should be listed on the incoming inspection report, together with the disposition. The photo department should take all of its instructions for setting up the working film from the incoming inspection report. The photo, screen printing, and inspection departments should be able to tell if they anticipate a problem on a given job by reading the notes and numbers on the incoming inspection report. Ideally, whenever a problem is noticed by shop personnel, they should be able to consult the incoming inspection report to see if it has already been questioned and dispositioned. The incoming inspection report is a basic document with which other shop personnel should become familiar.

Planning Repeat Jobs

In planning repeat jobs, the planner has another opportunity to help improve the yield and resolve problems. Hopefully, a well-planned repeat order will run through the shop more smoothly and with greater yield than the initial run. Part of the planning procedure for repeat jobs must be a review of all available documentation on the prior run or runs. This involves a review of previous traveler(s), including (1) preshipping inspection forms and (2) notes added to the incoming inspection report which ran inside the traveler of previous runs.

A review of previous travelers, and other forms running with them, will shed light on problems which occurred on previous runs. The planner should look for the following information:

1. Have notes been made about situations which must be corrected on subsequent runs?
2. Was the yield dropout particularly high on any of the operations?
3. Are any reasons listed for scrap or required rework?

4. Was the previous run shipped on time?
5. Was the previous run shipped completely?
6. Does the previous traveler show any overage going into stock that can be used on the current run?
7. Is the price on the current run in line with prices received in the past?

The alert planner will attempt to learn everything about the prior run before releasing a subsequent job. Sometimes jobs have several travelers when the part number is ordered. Later releases are placed in a file for release at the appropriate time. The planner must not pull travelers from the later release file without investigating the prior runs. If an earlier run has not yet been shipped, the planner should obtain information on that run. There may be valuable information noted on the traveler which should be incorporated into the later releases.

Dealing with Other Departments in the Shop

The planner will deal with other supervisors to discuss shop capabilities, setting up jobs, making revision changes, and making inhouse changes such as drill sizes and panel layout. From time to time, the planner will issue instructions needed to address certain situations, like those listed above. The planner should set up a feedback or follow-up system to ensure that the correct action was taken and in a timely fashion, before the job is processed. Written memos to the appropriate supervisor (and manufacturing manager) are one way to ensure correct and timely action. The incoming inspector can be of service in following up manufacturing instructions issued by the planner.

Dealing with Customers

Customers are like anyone else; they like doing business with vendors who are easy to deal with. They do not like getting repeat telephone calls about problems. However, sometimes there is no choice but to call the customer for a disposition. This is the case with design errors and when inadequate information has been supplied to do the job correctly. When the planning and incoming inspection functions have been completed, the planner should make every attempt to disposition items which have been questioned. No call should ever be made until all information has been reviewed, including the customer's procurement specification and the spec summary sheet. The spec summary sheet can be used to note information about the customer, such as routine waivers, and any other items. Even when the customer has no procurement specification, the planner can initiate a spec summary sheet. On this sheet can be placed information about the customer for future reference.

Whenever the customer is called, the call forces the buyer or engineer to stop work and obtain information to answer the planner's question. This can sometimes cause conflicts within the customer's company. Buyers and engineers sometimes come into conflict. Either individual may use repeated calls from the board vender as a personal defense and as ammunition to attack the other. The vendor should stay out of these situations.

There are times when a board is not manufacturable to the customer's specification—for example, when tight pad-to-trace air gaps cause a solder mask problem. Or the customer may require no mask on the pad and no exposed traces, yet the artwork has only .006 inch spacing. Sometimes a customer is willing to listen, and sometimes not. However, whenever a manufacturability problem is detected, it must be addressed. Before calling the customer, it is wise to make a list of what must be communicated, as well as a summary of how the planner will respond to the customer's likely response. Before calling, ask yourself the following questions and be able to explain or defend the answers to the customer:

1. What is the exact condition causing the problem? Does it involve the artwork, blueprint, spec requirement, etc.?
2. Why does this condition cause a manufacturability problem?
3. How can I explain this simply and clearly to the customer? (Try writing down your explanation.)
4. What solutions can I use inhouse to help the customer and save time?
5. Will any of these possible solutions cause other problems? (If so, you must be prepared to explain them to the customer.)
6. What action can the customer take to solve this problem for me if I cannot solve it inhouse?

If, for instance, pad-to-trace spacing is .006 inch, you can be certain that either traces must be exposed by solder mask or solder mask must be allowed on the pads. Most customers will allow mask on the pads in preference to exposed traces. However, if your shop is responsible for electrically testing the boards, you must know if allowing mask on pads will cause testing problems. If the holes being tested have only a .002-inch annular ring, you cannot allow the customer to approve solder mask on the pads. If the boards do not test well, they are scrap. Your company has just lost a lot of money, and the customer still does not have good printed circuits. Nothing is gained by agreeing to something that cannot be done.

When explaining problems such as the common one above, you must make sure that the customer understands the problem and the result of taking agreed-upon action. If the customer will not change an unrealistic tolerance or other requirement, you must consider refusing the job. There is no point in expending time and money on a requirement which cannot be met and to which you are committing your shop's resources and reputation. If the planner comes across a printed circuit with design requirements exceeding those listed in IPC and military specifications and standards, beware. If the artwork exceeds the precessing and tolerance requirements listed in this book, beware. If a customer will not change a requirement which cannot be met, beware. Review the situation with the managers of your company. As a planner, you should not commit your shop to requirements which are unrealistic.

There are times when a decision will be made by the shop's management to proceed with a job. The reasons usually are as follows: (1) The customer probably does not inspect for and will not notice that the requirement has not been met; (2) management has decided to build the boards and ship them to the customer. If the customer accepts them, fine. If the customer rejects them, management is prepared to disqualify that customer and never do business with him/her again. Some customers will leave you no option but to turn the job down or proceed with it as a gamble. At least you should allow your managers to proceed, knowing exactly what the possibilities are.

Tin-Nickel/Flat Mask and SMOBC Processes

Both of these processes are becoming more popular. Both of them offer an aesthetically appealing board after wave soldering has been performed. The most difficult problem facing the printed circuit manufacturer today is applying solder mask without exposing traces or getting mask on pads. The planner may not be aware of this fact, but the people in the final inspection department are, and so is the quality assurance manager. This problem is magnified many times by the solder mask over bare copper process.

BASIC OPERATIONS

The SMOBC process exposes the boards to several processing conditions which should be avoided:

1. Stripping of tin-lead: the dielectric material (laminate) is exposed to powerful acids and oxidizers. There is a brief period (1 to 2 minutes) during which the tin-lead must be stripped and the panels immediately removed and rinsed; otherwise noticeable degradation of the butter coat, measles, and crazing will occur. Chemical attack on the laminate causes degradation of the dielectric material even during this brief period. However, the attack is not noticeable to the unaided eye if exposure is limited to this 1 to 2 minutes.

SMOBC	Tin-Nickel/Flat Mask
1. Drill panels	1. Drill panels
2. Plate through holes	2. Plate through holes
3. Image with plating resist	3. Image with plating resist
4. Electroplate: copper, solder	4. Electroplate: copper, tin-nickel
5. Strip resist/etch copper	5. Apply flat mask resist
6. Chemically etch solder from copper circuitry	6. Electroplate solder
7. Inspect panels for complete solder removal	7. Strip all resist/etch copper
8. Plate contact fingers	8. Plate contact fingers
9. Apply solder mask	9. Reflow plated solder
10. Dip panels in molten solder	10. Apply solder mask
11. Strip SMOBC tape from contact fingers; remove tape residue	
12. Touch up solder mask or rework contact fingers where tape leakage occurred.	

NOTE: It does not matter whether contact fingers are plated before or after solder masking. If they are plated afterward, other problems occur, such as lifting the mask.

It is time to ask: How can dielectric material which has been exposed to some of the most powerful acids and oxidizers known produce a superior printed circuit?

2. Immersing the panel in molten solder (hot air leveling) is very similar to performing thermal stress (solder float). Thermal stress is typically performed on quality control coupons prior to microsectioning. This is considered the acid test and can result in laminate voids, resin recession from the hole wall, copper separation from the hole wall, cracked copper in the barrel and at the knee of the hole, pad lifting, and other undesirable conditions. For these reasons, dipping printed circuit panels in molten metal should be reserved for wave soldering.

It is almost always necessary to run panels repeatedly through molten solder dipping operations, due to plugged holes, funneled holes, exposed copper in holes, exposed copper on pads due to solder masking limitations or bleed, and so on.

It is time to ask: How can thermal shocking of a printed circuit prior to assembly produce a superior printed circuit?

3. Prior to immersing printed circuits in molten solder, a special tape must be applied to prevent the contact fingers from being coated with solder. Anyone who has been involved with hot air or hot oil leveling has seen what happens when this tape leaks. Solder flows onto the contact fingers. This is a common problem with SMOBC and causes much scrap and rework for the board

manufacturer. Even if the copper contact fingers have not yet been plated, the tape leaks and solder covers part of the finger surface. This condition is actually worse than getting solder on the gold, since it is virtually impossible to remove it from the copper without chemical attack of the copper and laminate. Also, the hot air leveling tape can pull solder mask from the conductors. This tape has another drawback: It leaves a heavy residue on the laminate between the contact fingers. This residue cannot be removed easily.

It is time to ask: How can a manufacturing process (unique to SMOBC) which results in rework and scrap for the board manufacturer be economically more sound than using a method which avoids this costly problem? Also, how can printed circuits with a heavy tape residue between the contact fingers be more reliable than a printed circuit with clean laminate at the contact fingers?

4. Independent operators of solder dipping equipment (hot air/hot oil leveling) must run the solder as long as possible. Typically, the solder is run until the deposit is so severely dewetted and grainy that the aesthetic appearance of the boards is very poor. The level of copper content in the hot air leveling solder is well above 200 ppm, often 400 to 800 ppm at this point. This is why SMOBC printed circuits almosts always have dewetted solder. The greater the area covered by solder, the more severe the dewetting.

It is time to ask: How can severely dewetted solder, and solder with 200 to 800 ppm of copper, make a more reliable printed circuit with more reliable connections?

5. The dewetted solder of SMOBC poses further problems for surface mount printed circuits. The solder on the pads is very uneven. The solder at the crest of the puddle can often reach .002 to .003 inch, and on the same pads, the solder can be too thin to read by microsectioning. The thin portion of the dewetted pad also impedes) solderability. The thin solder will not protect the copper of the pad from oxidation. If the board is set on a shelf for an extended period, there is little to protect the solderability.

It is time to ask: How can a surface mount board with uneven, dewetted solder puddles at the pads provide an optimum surface for part placement and solderability?

The conditions listed above are unique to the SMOBC process. By using the tin-nickel/flat mask process, all of these undesirable conditions are avoided.

The tin-nickel/flat mask process follows the natural order of plating. There is no need to strip metal chemically; and then to reapply that same metal by immersion in molten solder. The process is as straightforward as that of traditional printed circuit manufacturing. There are no operations which are virtually guaranteed to produce scrap, rework, degradation of the laminate, plated-through holes, and surface mount pads.

Copper tends to build up in both the molten solder of hot air/hot oil leveling equipment and in the tin-lead plating bath. This situation is natural and unavoidable. However, the level of copper can be kept under 20 ppm (dewetting typically becomes noticeable at 60 to 100 ppm of dissolved copper) by normal dummy plating. Copper plates out of the tin-lead bath quickly at low-current density plating (3 to 5 amps/ft^2). Copper cannot be removed from molten solder. It continues to build up until the operator is forced to dump the solder and replace it with fresh solder. Given a solder cost of $1100 to $1200 and a lot of down time, this is obviously an expensive operation, and one which will be postponed as long as possible.

Tin-Nickel/Flat Mask: Superior Quality

1. The tin-nickel/flat mask process places a barrier between the copper and the solder. The copper cannot migrate into the solder, and air cannot oxidize the copper. As a result, solderability is enhanced, and shelf life and usability are extended indefinitely.
2. Insignificant amounts of copper in the solder. The only economically practical method of applying solder which can eliminate dewetting is by using a well-maintained solder electroplating bath.
3. No exposed copper on the pads. Solder already exists on the pads and in the holes prior to solder mask application.
4. Enhanced reliability of the dielectric material. There is no need for the laminate to be exposed to strong acids and oxidizers.
5. Enhanced reliability of the plated-through hole wall. There is no need for repeated thermal stressing of the printed circuit panel during the manufacturing process.
6. Controlled thickness of solder. The solder thickness can be precisely controlled by electroplating. There is no reliable way to control the thickness of solder which is applied by immersing the printed circuit in molten metal and then blowing it from the holes and circuits with hot air or hot oil. Even the pressure of the hot air will not control unevenness from dewetting.
7. Pure solder coats the holes and pads for assembly operations. The nonactivated flux is all that is required for wave or drag soldering operations. Solderability is preserved and enhanced by this process.
8. It is possible to run with tighter circuit air gaps with the tin-nickel/flat mask process using traditional solder masking methods. There is no need to resort to costly dry film solder mask to ensure that traces are covered when they are closely positioned to pads.

Chapter 6

Planning Flex and Rigid-Flex Jobs

WHY ARE FLEX CIRCUITS MORE DIFFICULT TO MANUFACTURE THAN RIGID CIRCUITS?

Companies interested in manufacturing flex and rigid-flex circuits should keep in mind that there is no magic formula for manufacturing good boards. There are only more operations, added precautions, and the need for more extensively thoughtout planning than is required for traditional rigid multilayer boards. As with any aspect of printed circuit manufacturing, there is a lot to be learned simply by observing the results and learning from them. There are some basic facts about flex circuits of which designers, buyers, and quality assurance, planning, and sales/quotation people should be aware. Most of the problems with flex circuits are caused by a few peculiarities of the materials used to make them. The most common materials, and the only ones which will be discussed in this book, are Kapton dielectric with acrylic adhesive.

PECULIARITIES OF FLEX MATERIALS

1. Kapton is dimensionally unstable; this creates registration problems that must be overcome. Kapton copper-clad material tends to shrink as the copper is etched off. This shrinking creates registration problems from one operation to another and frequently results in the need to compensate drilling programs and artwork. Fortunately, shrinkage occurs only in one direction and can be controlled. Once shop personnel become accustomed to shrinkage, dealing with it becomes routine. Once a job has been run, and any necessary compensations made to the drilling programs and artwork, future releases will run in an identical manner. However, the tendency for Kapton material to shrink during processing is probably the most important difference between manufacturing flex circuits and manufacturing rigid circuits. Kapton shrinks about .001 inch for every inch of panel length if there is absolutely no copper left on the panel. The more copper on the panel, the less the shrinkage.

Copper can be added in border areas and between circuits. With a solid 1-inch border and metal between images, shrinkage is cut to about .0007 or .0008 inch per inch of panel length. If tolerances are not too restrictive, it is possible that no compensation will be necessary for many jobs. It is important to keep as much copper as possible on flex layers; this is a more important requirement for flex layers than for rigid layers.

2. Coverlay material, the Kapton cover for outer circuitry, also changes dimensionally. Prior to punching or drilling holes into it, it must be stressed relieved. This is relatively simple. The polyolefin protective cover is simply peeled back from the adhesive surface; this allows the coverlay to shrink slightly. After a few minutes, it can be easily reapplied. Once this has been done, the coverlay sheets can be punched (tooled) or drilled.

3. Kapton absorbs moisture. It grows dimensionally with moisture absorption and shrinks as the moisture is removed. Thus, steps must be taken to bake it and keep it dry in order to void further dimensional problems. Remember, Kapton is a polyimide, all polyimides tend to absorb moisture. Since the adhesive surface is acrylic, as opposed to polyimide prepreg, the moisture absorption problem is not severe. There is no major moisture absorption problem with acrylic adhesives.

4. Kapton materials are very thin. This creates severe handling problems which must be overcome whenever pieces of Kapton are cleaned, baked, exposed, developed, etched, touched up/inspected, or moved. For instance, when etching flex layers, it is common to place them inside a frame. The frame can be made from FR-4 laminate. All it must do is hold the edges flat during processing. Flex layers do not go through most scrubber/driers well; therefore, it may be necessary to scrub manually and pat dry using newsprint. Many considerations like these must be kept in mind when handling Kapton flex layers.

5. Acrylic adhesive does not flow during lamination. Therefore, it does not remove air in the same manner as a prepreg, such as epoxy. To help force air out during lamination, a conformal coating is used. One of the best conformal coatings is 20-mil-thick plastic, available in vinyl and other materials. The conformal coating is separated from the panel by a release sheet, such as Pacothane or Tedlar. The conformal coating is pressed down and around the copper circuitry during lamination, forcing air out. It may be difficult to remove all air during lamination, and sometimes other methods must be used. Higher pressure is helpful. Reducing the copper on the panel, to provide more relief between boards and along boarders, will help keep

trapped air outside of the fab lines. If air remains inside the fab lines after lamination, it can still be removed. Simply make small cuts in the coverlay sheet along the circuitry where the air is trapped; then relaminate the boards using conformal coating. Air will escape through the sliced coverlay, and the adhesive will seal the slices. Note that this technique cannot be used on boards being built to MIL-P-50884C; however, MIL-P-50884C does allow some air entrapment.

6. Flex materials are more delicate than materials for rigid printed circuits. They are thinner, softer, and more sensitive to thermal, mechanical, and environmental handling and processing. Epoxy, for instance, is a thermo-setting plastic. Once cured, it remains cured. Acrylic adhesive is a thermoplastic. It does not remain strong and fully cured at elevated temperatures. Acrylic adhesive can be subjected to multiple lamination operations, if necessary, with no adverse effects.

All flex circuits, including single- and double-sided boards, must be laminated in a press.

The circuitry must be covered by Kapton. Thus, even a single- or double-sided board must go through a lamination operation. A coverlay sheet of Kapton must be drilled, registered, and laminated to the circuitry. Since the panel has probably been reduced in size after the copper has been etched, it may be necessary to adjust the drilling program before drilling holes in the coverlay sheet. As with all lamination operations for flex circuits, conformal coating must be used on both sides of the panel to force air out. The adhesive does not flow; this cannot be forgotten.

WHY ARE RIGID-FLEX CIRCUITS DIFFICULT TO MANUFACTURE?

A typical rigid-flex circuit is like a cable with two printed circuits connected to it, one at each end. The connection is made by laminating the cable between pieces of rigid laminate. After lamination, holes are drilled and plated to make electrical connections to the cable. The side of the rigid laminate bonded to the cable may have circuitry etched into it before it is laminated. The outer side of the rigid laminate will eventually have circuitry, or simply pads around the holes. It is the plated holes which connect the printed circuits through the cable. Instead of being a cable, the flexible part actually consists of layers of circuitry on flexible Kapton. It is common to refer to the flexible portion of a rigid-flex circuit as a "cable."

How is the rigid-flex circuit produced (see Figure 6-1)? The cable, which is actually the flex layers, is laminated between two sheets of rigid laminate, which already has circuitry etched on the inner layer side. So far, this is just like

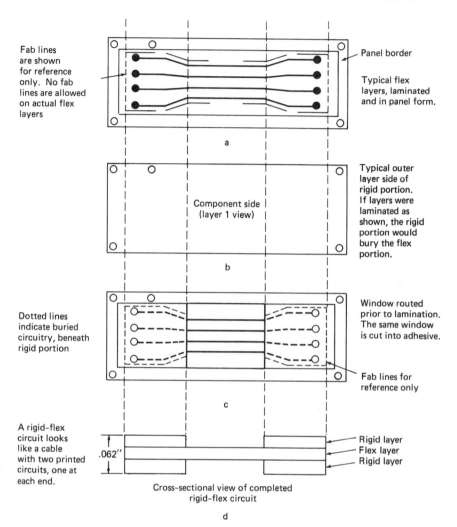

Fab lines
are shown
for reference
only. No fab
lines are allowed
on actual flex
layers

Panel border

Typical flex
layers, laminated
and in panel form.

a

Typical outer
layer side of
rigid portion.
If layers were
laminated as
shown, the rigid
portion would
bury the flex
portion.

Component side
(layer 1 view)

b

Dotted lines
indicate buried
circuitry, beneath
rigid portion

Window routed
prior to lamination.
The same window
is cut into adhesive.

Fab lines for
reference only

c

A rigid–flex
circuit looks
like a cable
with two printed
circuits, one at
each end.

.062″

Rigid layer
Flex layer
Rigid layer

Cross-sectional view of completed
rigid–flex circuit

d

Fig. 6-1   Construction of a rigid-flex multilayer.

making a traditional rigid multilayer circuit. This rigid laminate will be used to form the rigid portion of the rigid-flex. There is just one problem: If the flex layers are laminated between the rigid laminate, how can they flex? Also, rigid laminate is needed only at the ends of the cable, not along its entire length. Since the rigid portion is desired only at the ends of the flex layers, all of the rigid laminate which covers the flex layers must be routed away. This brings up another problem: If the flex layers are laminated between the rigid laminate, how can the rigid laminate be routed away without damaging the flex layers?

The only way to end up with rigid laminate at the ends of the cable (flex layers), and avoid sandwiching the full cable, is to partially route the rigid laminate prior to laminating it to the flex layers. Thus, window-like openings will have been routed out of the rigid layers just before they are laid up on caul plates with the flex layers for lamination pressing.

The adhesive used to bond the flex and rigid layers will also need windows to be cut out. The windows in the adhesive must match the windows in the rigid layers. The panel is now processed in a manner similar to that of traditional multilayer printed circuits. The window in the panel causes some imaging, touchup, and plating difficulties. The final fabrication for board outline is a two-part operation. Only the rigid part of the rigid-flex panel can be routed. After this is done, the flex portion of the panel will have to be cut manually, usually with an Exacto-type knife.

To summarize: Rigid-flex multilayers are more complicated to make than other types of circuits for the following reasons:

1. A flex circuit (cable) must be manufactured. (Make the flex layers.)
2. Rigid layers must be manufactured with circuitry only on one side—the inside—which is laminated against the flex cable.
3. Rigid layers must have windows routed into them; the windows expose the body of the flex multilayer cable. The rigid portion must have the windows cut now; they cannot be cut after lamination without damaging the cable.
4. The adhesive used to bond the layers must have windows routed into it which match the windows in the rigid layers.
5. The rigid and flex layers are bonded in a lamination press cycle similar to that of traditional rigid multilayer circuits. Conformal coating must be used to help assure air removal, since acrylic adhesive is used for bonding.

   NOTE: The laminate plug which was routed out to form the window in the rigid layers is now placed back inside the window for the lamination operation. This keeps the pressure even and avoids damaging the flex circuitry. Since the adhesive has an identical window cut into it, the "window plug" can be easily popped out of the panel after lamination. When the plug is removed, the panel looks like an ordinary multilayer panel, except that it has a window covered with flex circuitry.

6. The panel is now ready for drilling and electroless copper deposition to plate through the holes.

7. After electroless copper application, the panel is ready for outer layer imaging, just like a traditional multilayer printed circuit.

   NOTE: There is a depression in the panel where the window is covered by the flex circuitry. Special touchup is required to assure that this window depression is completely covered by resist, including the edges of the depression. This is necessary to keep the edges of the window from plating.

8. After plating, resist stripping, and etching, the panel is ready for fabrication of final board outlines.

NOTE: The remaining portions of the rigid layers are now routed from the panel. Note that this is the second routing operation. The board is now connected to the panel only at the panel borders of the flex layers. Since Kapton is difficult to route, the final board outline will have to be completed by cutting the Kapton with an Exacto-type knife. It is possible to help make Kapton routing easier by interleafing release sheet material (Tedlar/Pacothane) between all layers, including multiple flex layers. This allows compressing, which will hold Kapton in place long enough to route. If only small cuts are required, it may be easier to simply cut the Kapton with a razor knife. Two N/C routing operations have been performed; a manual cutting operation is still required to complete the board outline.

ENGINEERING REVIEW

The customer who designs the flex circuit may not be an expert. Even MIL-STD-2118 is a fairly new document. Prior to its publication in 1984, little information on designing flex circuits was available. Even less has been available on manufacturing flex circuits. The process is mainly one of trial and error. For these reasons, the printed circuit manufacturer will usually find that customers are willing to work with them to overcome design and manufacturability difficulties. Also, customers will assume that flex circuit manufacturers know what they are doing and will expect a thorough engineering review of all documentation.

The first thing to do when faced with quoting on or manufacturing a flex or rigid-flex circuit is to obtain the answers to a number of questions. These answers are needed before the cost, difficulty, and required lead time for the job can be determined. Important questions include the following:

1. How many layers are there? Is the board type 1, 2, 3, 4, or 5? If it is type 3 or 4, what is the cost and lead time for performing plasma desmear?
   The number of tooling operations and the degree of difficulty rise quickly as the number of layers increases. The type 4 rigid-flex is far more complex to tool and build than the type 3 flex multilayer. Multilayer and rigid-flex boards will require plasma desmearing for reliable inner layer connections. When it is known that plasma desmearing is required, arrangements must be made for this service. If a given printed circuit shop has a plasma etcher, this is no problem. Most shops, however, have to contract this process to an outside service. Cost and timing considerations come into play. The

person providing the quotation will need firm cost numbers and an approximate lead time. The difference between 2 days and 7 days of lead time for this service can have a significant impact on the cost of the service and the amount which the board manufacturer can charge for the product.

2. Is assembly required?
   a. Will the customer supply parts,or will they have to be ordered?
   b. If parts must be ordered, What parts and quantities are involved?

3. What minimum order quantity does the supplier requirer? Sometimes the customer is willing to supply part kits for assembly. This reduces the cost and effort burden on the board manufacturer. Many flex circuit users ask that fully or partially assembled circuits be supplied to them, with the board manufacturer procuring the components. For the board manufacturer, even installing and soldering connectors can be a major cost consideration. A customer may want, for example, 10 to 100 boards which have a $3 connector soldered in. If the minimum purchase quantity to get the $3 price is 1000 pieces, the board manufacturer will have to think twice about this job. It is not unreasonable to ask the customer to either buy all 1000 connectors or sign a long-term contract for more than the 10 to 100 boards. The customer may be able to procure the smaller quantity through channels not available to the board manufacturer. When all is said and done, the board manufacturer who is serious about entering the flex circuit market may have to consider permanent arrangements for handling assembly requirements, since so many flex customers are demanding that assembly be included in any deal for flex circuits.

4. Are pins, housings, or hardware required? This is a special case of limited assembly being required. As with rigid, traditional printed circuits, it is common to install various types of pins or housings in flex circuits. Unless the board shop is experienced in installing these items in flex circuits, it may have neither the equipment nor the hardware required. The cost of the hardware item(s), availability, and lead time must be known before a quotation can be issued or the job released from the planning department. Planners should keep quotation personnel updated on hardware capability, costs, and availability requirements. Installation of various types of hardware may have to be done by outside contractors until the board manufacturer determines that he/she wants to perform this service inhouse.

5. Is a stiffener required?
   a. Will the stiffener be route and retain? (Yes, if soldering is done.).
   b. What is the cost and lead time for installing pins, housings, or hardware?

Stiffener requirements should be clearly specified on the blueprint. The stiffener is a rigid piece of laminate (GF or GI) which is bonded to the finished board. Sometimes 5-mil Kapton or Kapton tape is used. The stiffener supports the board at a connector or some other component. The requirement for a stiffener does not make the board a rigid-flex, type 4 multilayer. Even a single-sided board may require a stiffener.

The stiffener does increase the difficulty and cost of manufacture. It requires at least one drilling and one routing program, and at least one drilling and one routing operation. If the stiffener supports the circuit during assembly (wave soldering), it may require a route-and-retain pattern and can be fairly complex and time-consuming to program, drill, route, and/or bond. If the circuit will be exposed to continuous flexing, the stiffener requirements are slightly more complex. Stiffeners for such boards (class B of MIL-STD-2118) are chamfered or have a radius along the edge where the stiffener meets the flex portion of the circuit. This creates more work for the board manufacturer. Also, a bead of adhesive (epoxy or some other type) often must be applied at the stiffener along the chamfered or radiused edge. This creates even more work for the board manufacturer and must be included in the quotation. There is an increasingly common alternative today when a radius or chamfered edge is required with a bead of adhesive. Both edge processing and bead application can be avoided (saving time and money) by simply having the board manufacturer extend the bonding adhesive slightly out from the edge of the stiffener (see Figure 6-2). This will keep flex circuitry from being damaged by the stiffener during flexing.

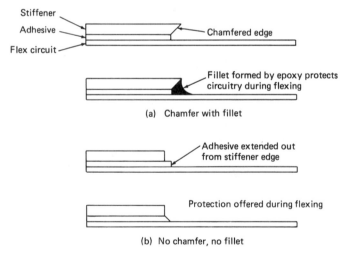

Fig. 6-2 Extending adhesive from rigid layers or stiffeners offers an alternative to chamfer and fillet.

6. Is electrical testing needed? What are the requirements? If electrical testing is required, a test fixture will probably have to be made. The test fixture setup is similar to that for rigid boards. However, it may be possible to test with a cable tester. If the flex circuit forms a ribbon or cable, the testing requirements can be simplified. Cable testing equipment is fairly inexpensive. Its use will facilitate testing after the board has had connectors soldered in or otherwise installed.

   If electrical testing is to be performed after assembly, this requirement must be made clear to the person doing the quoting and planning. This is a more sophisticated type of testing than most board shops are prepared to provide. It will be necessary to consult closely with the customer, and with another testing facility, if testing is required after assembly. Note that testing even after connector and housing installation is far more difficult than bare board testing.

7. Is the longest board outline dimension 24 inches or less (or whatever maximum dimension the shop press can handle)? It is common for flex circuits

   4, 36, 48, or more inches in length are
   ting orders for long circuits, it is neces-
   can handle. Generally, the longest circuit
   largest caul plate (lamination plate) the
   n be run, but changes in the process are
   lities. A long circuit can be made by a
   is case, part of the circuit will stick out
   a lamination sequence. Eventually, the
   ve been laminated. This type of lamina-
   moplastic characteristics of the adhesive.
   ot like an epoxy prepreg. Acrylic adhe-
   n be "remelted" with no problem. With
   -like state regardless of how many times
   only slightly longer than the press, it is
   stick out of the ends of the press. The
   mped shut. The chief limitation is that the
   the adhesive does not receive adequate
   consult your company's engineering and
   s of how the shop decides to handle the
   lt and time-consuming than short ones.

   quired for final board outline fabrication?
   accurate method of fabricating the final
   a die. Steel rule dies are fairly cheap (a
   le. The company which makes the die can

also punch the circuits unless the circuit shop desires to set up its own punch. Unless the quantities are very small and the board outline is very simple, it is best to use a steel rule die to obtain the final board outline. Kapton is difficult to route satisfactorily. It can, however, be cut with Exacto-type knives. Manual cutting for fabricating final board outline is time-consuming and not very accurate.

9. Are there any openings in coverlay sheets other than holes? Complicated openings in coverlay sheets make the manufacturing process more difficult. Since openings other than holes cannot be drilled or routed, they will have to be punched or cut manually. Regardless of which method is used, complicated openings in the coverlay sheets will add difficulty, time, and cost to board manufacture. Windows, slots, "square holes," and other openings which cannot be drilled should be punched with a steel rule die for ease, accuracy, and cleanliness of the cut.

10. How many of the following are required?
   a. Drill programs
   b. Drilling operations
   c. Fabrication programs
   d. Fabrication operations
   e. Lamination operations
   Even a simple single-sided (type 1) flex circuit usually requires two drill programs and two drilling operations. One drill program and drilling operation is for drilling the flex printed circuit. After imaging and etching, the panel has been reduced in size through shrinking. It has been reduced a maximum of .001 inch per inch of panel length as the copper has been etched away. If the coverlay sheet were drilled using the same program, the coverlay would not register correctly. To optimize registration of the coverlay sheet to the printed circuit panel, it is necessary to compensate the drill program.

11. How can the drill program be compensated? There are two methods for compensating the drill program. The first method is to resize the program to fit the shrunken panel. This can be accomplished by using the sizing option in the software of the N/C programming equipment. This function may be used for matching a drill program to dimensionally altered artwork in order to improve registration. The method works best for large boards, which run the length of the panel. The holes can be compressed in the direction where shrinkage occurred. If the etched panel has shrunk .00075 inch per inch of board length), the program should be compensated by this amount.

EXAMPLE: If a panel shrank .012 inch in the 18-inch lengthwise direction, it was reduced .012/18.000 = .000666 inch/inch. The drill program must therefore be compressed .012 inch. If the coverlay sheets are now drilled with this resized program, the coverlay access holes will register quite accurately to the holes in the flex panel.

The second method of compensating the drill program works best for small boards being run multi-up on a panel. The spacing between the boards can be reduced by the amount of shrinkage per inch.

EXAMPLE: If an 18-inch-long panel shrank .12 inch, it shrank .000666 inch/inch. If the boards are 2 inches long in the lengthwise direction, approximately .0012 inch of spacing must be removed between the boards in the lengthwise direction. Coverlay sheets drilled with such a program register quite accurately to the holes in the flex panel.

The need for more than one fabrication operation was discussed earlier in the section on rigid-flex multilayers. It should be kept in mind that manual completion of the board outline of the flex layers constitutes a third fabrication operation. It can even be argued that cutting windows into the adhesive is actually part of the fabrication operation. Also, rigid-flex multilayers require at least two lamination operations, one for generating the flex layers (the cable) and another for pressing the flex and rigid layers into a rigid-flex multilayer.

12. Are any shields or heat sinks required? Sometimes the blueprint calls for a special shield or heat sink to be applied over outer layer circuitry. In quoting and planning a job, it must be determined who will supply the shield or heat sink. Some shields and heat sinks can be applied easily. A soldering iron can be used effectively to tack a shield to a small area. Larger areas may require another lamination operation using the multilayer press. Thus, it must be decided not only who will supply the shield but also what steps are necessary to apply it.

13. Will 1-mil Kapton with 1-oz copper foil be adequate for all flex layers? One large manufacturer of flex materials recommends using 1-mil Kapton for almost all applications and 2-mil Kapton only if absolutely necessary. Due to the high cost of flex material, most board manufacturers stock only the material which will be most commonly used. If the blueprint calls for anything other than 1-mil Kapton, it is permissible to question the requirement as part of the planning and quoting process. Another consideration is that the thicker the material, the less flexible the board will be. Kapton with more than 1 or 2 oz copper is difficult to obtain on short notice. If a

circuit can be built on 1-mil Kapton, with 1-oz copper, this should be done. One thing the designer and planner should keep in mind is the adhesive used to bond the copper to the Kapton base material. Unlike epoxy/fiberglass laminate, copper-clad Kapton must have a layer of acrylic adhesive bonding the foil to the Kapton. The adhesive used is 1 mil thick. Thus, 1-mil Kapton which has copper foil bonded to both sides will have a dielectric spacing of .003 inch (.001 inch for the Kapton and .002 inch for the adhesive layers).

14. Plating: Copper with solder finish; copper/nickel/gold finish. Solder is the most common finish for flex circuitry. The solder is applied by hot air or hot oil leveling and is applied only to the holes. Even if tin-lead is plated as an etch resist over the copper, it must be removed. The tin-lead is removed because coverlay sheets must be applied to the outer layers of type 1, 2 and 3 circuits (single-sided, double-sided, and multilayer circuits). The coverlay sheets are applied only to bare copper. After coverlay lamination is completed, the panels are ready for hot air/hot oil solder leveling.

Double-sided and type 3 flex multilayer circuitry will have electroplated copper deposited on the copper foil. This is not an ideal situation for circuits exposed to continuous flexing. A shop which develops the technique for plating only the holes, and then applying a print and etch pattern, will have a definite competitive advantage over other shops. For a flex circuit which has plated-through holes, there is only one reliable method for avoiding plated metal in the areas which will be flexed continuously: Make the circuit a rigid-flex. Plating will be confined to the holes in the rigid portion.

Nickel/gold finishes should be reserved for outer layers of rigid-flex circuits (type 4). Unlike tin-lead, nickel cannot be removed easily from the copper conductors. Unless the shop has developed the technique for plating holes, followed by print and etch, nickel/gold finishes cannot be applied to type 1, 2, and 3 circuits.

15. Does the blueprint require the board manufacturer to prefold circuits? There is no need for the board manufacturer to prefold circuits. This is not recommended by MIL-STD-2118. Any such requirement should be reviewed with the customer and deleted. The board manufacturer may simply be opening the door for unnecessary problems.

16. What are the quality assurance and certification requirements?
    a. Final inspection only?
    b. Microsections required?
    c. Solder samples required?
    d. Certification to MIL-P-50884C?

(1) Is Group B required?

(2) Is Group C required?

(3) Is Class B flexibility required?

This book does not deal with the requirements of MIL-P-50884C. This specification requires that rather delicate flex and rigid-flex circuits be submitted to virtually identical testing requirements of MIL-P-55110D. If a shop has become certified to MIL-P-50884C, it is entitled to charge rather high prices for all circuits built to this specification. Some shops have not become certified to MIL-P-50884C, but they build boards which must pass the same tests. Before committing the resources and reputation of the company to these requirements, it is wise to perform inhouse tests to verify that they can be met routinely. Note that the B and C revisions of MIL-P-50884C are radically different. The B revision has no real requirements and does not call for a certification program like the C revision and MIL-P-55110D for rigid circuits.

Other items to beware of are the class B, continuous flexing requirement and class C flexibility endurance testing. Neither MIL-P-50884C nor MIL-STD-2118 recommends a class B requirement for anything but type 1 and 2 boards. Even the supplier certification testing requirements of this specification performs flexibility endurance testing only on type 1 circuits. No good can come from committing the company to a requirement which exceeds the industry standard, and it is unwise to agree to perform a job without fully understanding its requirements.

17. What quantity and date are required? These factors are considered because the quantity and lead time may be helpful in making important decisions. From the discussion so far, it should be clear that flex and rigid-flex circuits require many tooling operations, such as multiple drilling and routing programs and operations. Kapton does not N/C route easily; it must be punched or cut manually. Also, hardware and other component purchasing considerations come into play.

Small jobs are very expensive because of the tooling cost alone. Who can afford to tie up a programmer, a drill, a lamination press, and a router for the same minimum lot charge as a double-sided or rigid multilayer job? Flex and rigid-flex jobs take considerably more time to process. There are more tooling operations, manufacturing operations, and quality assurance requirements (including planning). All of these factors must be taken into consideration when preparing quotations and scheduling production. The shop has to understand the time and costs involved or risk losing a lot of money.

As an aid to understanding any given job, an engineering review form should be completed prior to giving a quotation. Completing an engineering review

ENGINEERING REVIEW: FLEX AND RIGID-FLEX

CUSTOMER_____ TODAY'S DATE _____

PART NO. _____ REVISION_____

BUYER'S NAME _____ PHONE _____

OTHER CONTACT _____ PHONE _____

QUANTITY DUE_____ DATE DUE_____

NO. LAYERS: FLEX_____ RIGID_____ BOARD TYPE: 1   2   3   4   5

BOARD DIMENSIONS_____ PANEL SIZE_____ NO. UP_____

PLASMA REQ'D: YES   NO   ,   OVERAGE THIS JOB_____

          : WHERE_____,LEAD TIME_____,COST_____

STIFFENER REQ'D: YES   NO   HEAT SINK REQUIRED: YES   NO

          : MATERIAL_____   MATERIAL_____

ASSEMBLY OR HARDWARE: YES   NO

          :   TYPE_____

          :   _____

          :   CUSTOMER TO SUPPLY   YES   NO

ELECTRICAL TESTING: YES   NO   SPECIAL REQUIREMENTS _____

STEEL RULE DIE - BOARD OUTLINE: YES   NO ,COST_____

          - COVERLAY/ADHESIVE YES   NO ,COST_____

NO. OF PROGRAMS/OPERATIONS: DRILLING_____

          : FABRICATION_____

          : LAMINATION_____

PLATING - COPPER/SOLDER: YES   NO ,   PRE-FOLDING REQUIRED: YES   NO

       - NICKEL/GOLD: YES   NO ,

QUALAITY - MIL-P-50884C: YES   NO ,   GROUP B OR C: YES   NO

       - MICROSECTIONS: YES   NO .   CLASS B FLEXING: YES   NO

       - CERTIFICATION: YES   NO ,   SOLDER SAMPLE: YES   NO

Fig. 6-3 Engineering review form for flex and rigid-flex jobs.

ENGINEERING REVIEW: FLEX AND RIGID

	MATERIAL TYPE	PANEL SIZE	NO. OF PANELS	PANEL SQ.FT.	TOTAL SQ.FT.	COST SQ.FT.	TOTAL COST
RIGID:_____							
RIGID:_____							
FLEX:_____							
FLEX:_____							
FLEX:_____							
ADHESIVE:___							
ADHESIVE:___							
COVERLAY:___							
COVERLAY:___							
CONFORMAL COATING:____							
HEAT SINK:_							
OTHER:_____							
:_____							

OTHER COST CONSIDERATIONS AND NOTES:
_____
_____
_____
_____

NONRECURRING TOOLING, INHOUSE CHARGES:_____

STEEL RULE DIE(S):_____

OUTSIDE PUNCH CHARGES:_____

COST PER PIECE:_____

ELECTRICAL TESTING:_____

INHOUSE LAB CHARGES:_____

OUTSIDE LAB CHARGES_____

PER PIECE COST_____

Fig. 6-3 (*Continued*)

form will also help the planner and other personnel understand the job more quickly (see Figure 6-3).

THE ROLE OF THE PLANNER IN COMMUNICATING REQUIREMENTS

The requirements for good incoming/artwork inspection are similar to those for rigid circuits. However, there are additional features, which will be discussed below. The need for good incoming inspection is even more critical. There are more tooling requirements, more manufacturing operations, and more quality assurance requirements. Furthermore, not everyone in the company may be familiar with the manufacturing of flex circuits. The planner cannot assume that other supervisors, managers, and engineers will fully understand all of the requirements simply by looking at a blueprint, with or without the traveler. The chances are better than even that the customer does not fully understand the design requirements and manufacturability either.

The planner must communicate much more closely with other shop personnel, asking and answering many questions. It is not uncommon for a planner to review each of the tooling requirements with programming, drilling, routing, fabrication, lamination, photo, and quality assurance personnel for every job. It is only when the planner has confidence in the established (written) procedures, and their meaning, that written instructions alone will be adequate. Even on some traditional rigid jobs, the planner must consult with other shop personnel. The need for this kind of communication in flex and rigid-flex manufacturing cannot be overemphasized.

PLANNING AND INCOMING/ARTWORK INSPECTION REQUIREMENTS

1. Customer Name, Part Number, and Revision Level. The information requirements here are identical to those discussed in prior chapters. Each piece of documentation—purchase order (sales order form), blueprints, and artwork—must have the same customer name, part number, and revision level. This applies to drill and magnetic tapes if they are supplied (see Figure 6-4).

2. Job Status. The requirements here are identical to those discussed in prior chapters. Each job must be checked to verify if that part number or revision level has been run previously. Careful investigation is necessary if the shop wishes to use any of the previous documentation. Any time previous documentation or tooling is used for a subsequent revision a risk is being taken. This decision should be a matter of company policy.

FLEX CIRCUIT: INCOMING INSPECTION

Customer	P/N	Rev.	New	New Rev
Date Inspected	Inspected By	Change	Repeat	
Film Number and Name	Drawing Number	Rev.	Neg.	Pos.

Item	Regstrn	Min. Trace	Trace/Trace	Airgap Pad/Pad	Pad/Trace	Artwork Neg/Pos	Comment
Layer 1							
Layer 2							
Layer 3							
Layer 4							
Layer 5							
Layer 6							
S/M: Component : Solder							
Legend: Component : Solder							

Hole Code	Size	Tol.	P/NP	Pad Size	Drill	A/R	I/L	Access Comp.	Access Solder	S/M
1) _____	____	____	____	_____	_____	____	____	_____	_____	___
2) _____	____	____	____	_____	_____	____	____	_____	_____	___
3) _____	____	____	____	_____	_____	____	____	_____	_____	___
4) _____	____	____	____	_____	_____	____	____	_____	_____	___
5) _____	____	____	____	_____	_____	____	____	_____	_____	___
6) _____	____	____	____	_____	_____	____	____	_____	_____	___
7) _____	____	____	____	_____	_____	____	____	_____	_____	___
9) _____	____	____	____	_____	_____	____	____	_____	_____	___

Fig. 6-4   Artwork/incoming inspection forms for flex and rigid-flex jobs.

1. Special Requirements
    -Are all copies of artwork present and usable without major modification?
    -Are all blueprints present and usable without major modification?
    -Is MIL-P-50884C a requirement of this job?
    -Group A or Group B required for this job?
    -Are coupons on the artwork, or are micro-sections a requirement for job?
    -Plasma desmear must be called out on traveler if coupons are a requirement
    -Electrical test required?
    -Any materials other than copper, Kapton, polyimide, FR-4, and acrylic?
    -Will panel length be greater than 24 inches, or part longer than 22 inches
    -Steel rule dies ordered for complex adhesion and fab patterns?
    -Symmetrical patterns clearly labled with alignment targets?
    -Layer sequence and side showing on blueprint?
    -Pads hitting traces?

2. Holes and Pads:
    -All pads have fillets?
    -All pads at least 2X hole diameter?
    -All nonplated hole pads captured. or with anchor spurs?
     --If not, is pad 2X + .010 over hole size?
    -Minimum annular ring: .015 single sided?
                         : .005 double sided?
                         : .002 inner layer?
    -All holes in coverlay sheet are round, no slots or irregular shapes?
    -Any nonfunctional pads?
    -Do any holes have access holes on one side of circuit only?
    -What metal(s) will be plated, if any, on pads?
    -Radii holes added, if hand fab called out?
    -Tear stops added for slots?
    -Die holes added, dimensioned from a datum?
    - Second drilled holes called out?
      -- Can second drilling be avoided?
    -Heat sinks or other unusual bonding requirement?
    -More than .002 inch adhesive required?
    -Bonding over more than 2 oz copper?

3. Hardware or Connectors
    -Hardware or connectors required this job?
    -P/N's: .   ------------------, ------------------, ------------------
    -Pins Sizes: ------------------, ------------------, ------------------
    -Hole Sizes: ------------------, ------------------, ------------------
    -Holes at least .008 inch diameter greater than pin diameter?
    -Stiffeners called out for connectors?
     --What material?
     --Will stiffener be adequate for wave/drag soldering?

4. Conductors
    -Bend Radius = 2 X Circuit Thickness : Thickness Circuit 1 _____
                                         : Thickness Circuit 2 _____
    -All conductor direction changes are curved, no sharp corners?
    -All conductors are perpendicular at circuit fold lines?
    -M/L Flex: no conductors on outside, pads only?
    -Rigid/Flex: S/M rep'd if conductors on outside of rigid?
    -Does film need to be compensated for trace reduction from etching?
      -- 1 oz foil?
      -- 2 oz foil?

Fig. 6-4 (*Continued*)

-Finished thickness 2 oz or less?
-Minimum airgaps .010 everywhere?
-When layers are registered, are conductors off-set from layer to layer?
-Minimum trace width .010?
-Minimum airgap .010?
-Ground or power planes required?
 --Cross hatched pattern?
 --Ground planes left unbonded in certain areas for flexibility?
-Any conductor thickness other than 1 oz called for?
 --Is material available without having to make it inhouse?

5. Plating Requirements
   -What metals are called out, and how thick?
        Copper_____   Tin-Lead_____
        Gold_____   Tin-Nickel_____
   -If nickel/gold called out, job must be rigid-flex.
   -Is selective solder deposition on pads required?
   --Hot air/hot oil leveling called out on traveler?

Problems, Dispositions, and Notes:

Fig. 6-4   (*Continued*)

Problems, Dispositions, and Notes:

Approval

Date

Fig. 6-4 (*Continued*)

3. Cataloging of Incoming Artwork. The requirements for knowing what was received from the customer are identical to requirements discussed in prior chapters. There must be a record of what was received from the customer.

4. The requirements for measuring the following features are identical to what has been discussed in prior chapter
   a. Layer-to-layer registration
   b. Minimum trace width, and tie bar width for thermally relieved power and ground plane connections
   c. Minimum air gap: trace to trace, pad to pad, and pad to trace
   d. Legend registration and minimum character width

5. Annular Ring and Inner Layer Clearance Measurements. These measurements are taken in the same way discussed in prior chapters, except that the laminate shrinks. The allowable fabrication tolerances in Table F-3 (MIL-STD-2118) should be consulted when the artwork inspection is completed. More "shop tolerance" will be required to compensate for material shrinkage.

6. Solder Mask Registration and Clearance Measurements and Requirements. These are almost identical to those discussed in prior chapters. Solder mask is required only on the rigid outer layers of rigid-flex circuitry. Furthermore, it is required only when there is circuitry; it is not required for pads alone. It is never to be applied to flex layers, and it is not required if the rigid layers have only hole pads, and no traces or other features. Any solder mask requirements besides those stated here must be questioned for disposition on the incoming/artwork inspection report.

7. Legend. Legend may be applied to flex and rigid layers.

8. Coverlay Sheet Access Hole Requirements. These are slightly different from those discussed in prior chapters (see Figure 6-5). Ideally, the access hole in the coverlay sheet will capture the pad (annular ring around plated or nonplated holes) for at least 270 degrees of the circumference (three-fourths of the circumference). It should be emphasized that pads on plated holes do not have to be captured. The requirement for coverlay capturing of pads, and for anchor spurs if there is no capture, applies only to nonplated holes. If the access holes do not capture the pads, the pads should have anchor spurs (see Figure 6-5B). Plated holes are anchored by the pads on opposite sides of the hole barrel. Also, traces which run into pads will anchor the pads. It is not permissible for nonplated holes to have pads which are not captured or anchored by the coverlay sheet. Coverlay sheets which capture

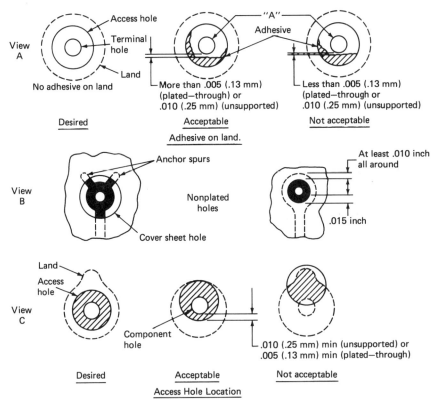

Fig. 6-5   Cover layer sheet requirements for plated and nonplated hole pads.

plated hole pads may cause minimum annular ring problems and may have to be evaluated for this condition.

The annular ring at plated and nonplated holes must be calculated and used in considering the size of the access hole. For component holes, MIL-STD-2118 (paragraph 5.5) recommends that the access hole be at least .030 inch larger than the diameter of the component hole. If it is .030 inch larger, there must be either a minimum of .010 inch overlap onto pads of nonplated holes or anchoring spurs on these pads.

No overlap or anchor spurs are required at plated-through holes. If the access hole does overlap, there must be at least .005 inch of exposed annular ring. Adhesive will often ooze from beneath the coverlay at the access hole. If adhesive does ooze, the amount of ooze will reduce the exposed annular ring. It is better if access holes have a clearance of at least .005 inch around the pads of plated holes.

Clearances for access holes are calculated in a manner similar to that used for calculating solder mask or inner layer clearance. However, for nonplated holes, coverlay should cover part of the pad; for plated holes, coverlay should either clear the pad entirely or allow for minimum annular ring requirements. If the clearance is less than the pad size, it should have a minus value (the pad is captured). If it is greater than the pad size, it should have a positive value. Minus values (captured pads) should be denoted by a minus $(-)$ sign in front of the value; positive values should have a plus $(+)$ sign in front of the value. This will alert the planner that pads are cleared by the access hole (desired for plated holes) or captured by the coverlay sheet (desired for nonplated holes). The basic formua for calculating access hole clearance is:

$$\text{Clearance} = A - B - C - D$$

$$A = \text{access diameter (drill size)}$$

$$B = \text{pad diameter (plated or nonplated hole)}$$

$$C = \text{front-to-back misregistration or artwork}$$

$$D = \text{shop tolerance (at last } \pm.004 \text{ inch)}$$

The shop tolerance is of major concern here. Table F-3 (MIL-STD-2118) allows at least .001 inch/inch for the standard fabrication allowance. This table does not, however, take into account that the drill program and artwork can be compensated during manufacturing. Each shop must determine its own shop tolerance. Initially, a value of at least $\pm.004$ inch should be used.

Obviously, negative values for access hole clearance (the pad is captured) should make the planner question the resulting annular ring for plated holes. Positive values for access hole clearance (the pad is not captured) on plated hole pads should make the planner verify that no traces will be exposed, just as with solder mask. A negative value at a plated hole pad should cause the planner to verify that a minimum of .005 inch of exposed metal will be open after oozing of adhesive. In such cases, it may be better to increase the size of the access hole. If the access hole clearance is positive for nonplated holes, the planner must verify that anchoring spurs exist or consider reducing the size of the access hole. If there are no anchoring spurs, the positive clearance is a discrepancy which should be questioned on the incoming report. An acceptable disposition would be to either reduce the access hole size or add anchor spurs. Adding anchor spurs would require an instruction to the photo department. If the access hole is reduced to capture the nonplated hole pad, it must be reduced enough to overlay onto the pad at least .010 inch after all registration factors are considered.

Sometimes the blueprint will call out access holes for the component side as well as the solder side. These must be evaluated separately.

9. Hole Charts. Several different drill programs and drilling operations are required in manufacturing flex circuits. The planner must study the blueprints to make sure that all required holes have been specified. There are some types of holes which the planner will have to add to the blueprint, such as radius holes (manual fabrication), die holes (steel rule die punching for fabricating board outline), tear stop holes, and any other tooling holes deemed necessary. Following is a list of most of the types of hole considerations that the planner must make sure are clear:

a. Punching or drilling of tooling holes (see Figure 6-6).

b. Primary drilled holes in type 1, 2, or 3 circuits prior to coverlay sheet lamination. If the job will have a board outline punched, die holes must be added to the blueprint. Die holes are .250-inch-diameter holes which are strategically placed about .250 inch outside the fab lines. They must be primary drilled and will be nonplated. They must also be drilled into the coverlay sheet. When this is done, they can be drilled greatly oversized. They should be drilled oversize so that misregistration will not reduce the effective size of the hole. These holes must be dimensioned from a datum on the blueprint. The planner must add these holes to the code list on the hole chart.

c. Drilling access holes in the coverlay sheets. Note that component and solder sides may have different numbers of holes and different hole sizes. If die, tear stop, and radius holes have been added to the drill chart for primary drilling, these holes must also be added to the drill chart for the coverlay sheet. When they are added to the sheet for coverlay drilling, the planner must make sure that they are drilled out far over the size used for primary drilling (at least .010 inch diameter).

d. Primary drilling in rigid-flex circuits.

e. Second drill operations. (It is good to avoid second drilling if possible, as this is complicated by registration problems due to the tendency of etched Kapton to shrink.)

f. Drilling of radius holes for flex circuits that will be cut manually. The planner should have redlined the size and location of radius holes onto the blueprint. Radius holes go at all inside radii, including notches and wide slots. Although the planner locates these holes by redlining the blueprint, the programmer must locate them with some degree of accuracy. These holes exist to facilitate manual cutting of flex circuits. The radius of the holes should extend .005 inch maximum inside the fab lines. The rings of the programming scope must be used to place the radius holes accurately. The programmer may not be accustomed to

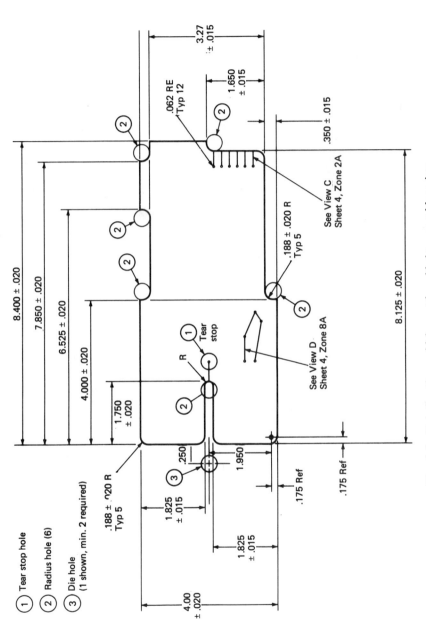

Fig. 6-6 Hole locations which must be added to most blueprints.

using the rings for this purpose. The planner must explain the requirement to the programmer. The person cutting the circuit will only have
to cut up to the radius hole. There will be no need to cut a radius manually—which is almost guaranteed to damage the circuit.

g. Drilling of tear stop holes at slots and slits in the flex circuit. The planner should have redlined the size and location of these holes onto the
blueprint.

h. Drilling of stiffeners. The hole sizes may have been called out on the
blueprint. If not, the planner must assign a hole and drill size and write
a drill chart for the stiffener on the blueprint. The holes must be drilled
out sufficiently over the corresponding hole in the printed circuit so that
any misregistration will not cause a problem when components are inserted. The only purpose of the stiffener is to support the board during
wave soldering and assembly.

i. Drilling of electrical test fixtures. For every drill program, there must
be a hole chart written on the blueprint. If the customer has not already
done this, the planner must look at the blueprint and all hole requirements. He/she should then draw up a chart and label it for its purpose:
primary drilling, coverlay sheets, stiffener, etc. These requirements must
be met before the job goes to the drill programming department. The
incoming/artwork inspector must verify that all of this work has been
done correctly.

10. Other Items on the Blueprint and Artwork
    a. Is MIL-P-50884C listed as a requirement on the blueprint? Is it listed
    on the other documentation? If so, do the artwork and blueprints meet
    the requirements of MIL-STD-2118? Is the shop management aware of
    the MIL-P-50884C requirement and its implications?

    b. Has planning called out for plasma desmear if the job is a multilayer
    flex or rigid-flex?

    c. Are the layer sequence and component sides clearly identified on the
    blueprint? Is each piece of artwork clearly labeled in terms of its layer?

    d. Is the hole pattern symmetrical? Have special notes of caution been
    redlined onto the traveler, incoming inspection report, and blueprint?

    e. Coverlay openings which are not round (holes) will require special in-
    structions on how they are to be cut or punched into the coverlay.

    f. Copper dams should be placed at slots, notches, slits, and other inside
    radii. These should be inside the minimum edge spacing requirement
    (see Table F-3 of MIL-STD-2118). These will help avoid problems and
    act as a backup to the tear stop holes. If copper dams need to be added,
    this instruction must be issued to the photo department.

    g. Bend lines must be perpendicular to circuitry.

h. If there are very few conductors at a bend, copper strips may have to be added. If so, the planner should issue this instruction to the photo department.

i. Are there any sharp bends in the flex area that need a radius? Sharp conductor turns should have a radius.

j. Is nickel/gold called for in the flex portion? This cannot be allowed and must be questioned on the inspection report.

Final note: When planning for flex and rigid-flex manufacturing, nothing should be left to an operator's interpretation. Planners must assume that other people will not understand the requirements. If this is assumed, the planner will review all requirements with other supervisors and engineers. After all, if the instructions are not clear to the planner, no one else can be expected to understand what has to be done. All manufacturing instructions must be on the traveler, blueprint, and incoming/artwork inspection report.

Chapter 7

Aspects of Quality Assurance

ASPECTS OF QUALITY ASSURANCE

Summary of Topics Covered to This Point

The purpose of this book is to provide the reader with a full understanding of what is required to manufacture a quality printed circuit. This is defined as a circuit which conforms to all specifications: blueprints, artwork, procurement specification, purchase order, and industry standards.

Standards and Specifications. In order to build a printed circuit which conforms to specifications, it is necessary to understand what the industry considers manufacturable. Without this knowledge, the designer, manufacturing engineer, or quality assurance person is not fully aware of what is happening. Without a knowledge of industry standards, attempting to design or build a printed circuit is like attempting to set up an airline schedule without knowing how long it takes to fly from one city to another. Without a knowledge of industry standards, there is no frame of reference for determining whether or not a requirement is reasonable and can be complied with. A detailed summary of common industry standards is presented in Appendixes A to F.

Understanding Blueprints. This section is not designed to turn the reader into a draftsman; its purpose is to provide a road map so that the reader understands common items on printed circuit blueprints. The manufacturing engineer and quality assurance person should be able to look at a blueprint and interpret it quickly and easily. When any items are not clear, it is easy to miss important information.

Processes and Tolerances. This information is presented for two reasons. The first reason is to help the reader understand why there must be tolerances on all dimensions of a blueprint and what the consequences of those tolerances are. There is a big difference between a $\pm.003$-inch tolerance on a plated holes diameter and a $+.002$ inch tolerance. The second reason is to help the manufacturing engineer manipulate artwork, materials, and processes to achieve the desired result. The manufacturing engineer cannot issue correct and thorough instructions to the shop floor without understanding the effect of each process on the product.

Artwork Inspection.   This section provides the manufacturing engineer with the means of inspecting artwork and determining its suitability for use—and, more importantly, for use in meeting specifications. The information presented must be used together with the information in Chapter 5.

Planning.   This section covers all of the duties of the person who must issue manufacturing instructions to the rest of the shop for each job. It is the planning engineer who is responsible for monitoring the accuracy and manufacturability of all documentation. It is also the planning engineer who must disposition artwork inspection reports and issue instructions to photo, drill programming, and other departments on how to set up jobs for the rest of the shop.

The planning engineer must also know when the documentation provided is incomplete, inaccurate, contradictory, or otherwise unsuitable for manufacturing to specification. This section shows how to use the information presented in the previous chapters. The planning engineer must be able to know how well a job will run in the shop.

Flex and Rigid-Flex Circuits.   This section discusses what makes flex and rigid-flex circuits different from rigid circuits. Peculiarities in materials, drilling, imaging, lamination, artwork processing, and other factors are discussed. Full instructions on performing an artwork/engineering review and dispositioning of that review are presented.

Quality Assurance and Inspection

One common definition of quality in common use today is ''conformance to specification.'' A product which conforms to all required specifications is said to be a quality product. The purpose of this book is to provide design, manufacturing, and quality assurance personnel with the background to achieve conformity to specification and to know when this cannot be achieved. Any person who is responsible for designing, inspecting, or building printed circuits can follow the principles presented in this book and know what will be required to manufacture the given design. This is not a book on how to perform each manufacturing operation. Detailed information on these operations is provided in the *Handbook of Printed Circuit Manufacturing*.

In other words, this book deals with engineering quality into the printed circuit. When this is done, the task of inspection proceeds more quickly and easily. This final chapter will discuss some of the inspection operations which can be performed during the manufacturing process to assure that quality is being built into the well-engineered product.

Incoming Laminate.   Laminate testing can be divided into two basic categories. The first consists of basic tests/examinations needed to verify that the material delivered is what was ordered and is accurately labeled. The second category involves laboratory testing to verify that the laminate passes the tests for the specification to which it is certified. These categories may appear similar, but the second requires much more sophisticated and time-consuming effort than the first.

Laminate inspection is a quick check to verify foil cladding, dielectric thickness, and visual appearance. It can be done rapidly and will uncover about 99% of the defects which concern printed circuit manufacturers. See Figure 7-1 for an example of a typical laminate inspection form. The planning engineer must have confidence that the laminate called out on the shop traveler is actually .008 or .022 inch, and not a thick .006 inch or a thin .028 inch. Laminate is not a complicated product. There is no reason to accept more than +.001 inch deviation in dielectric thickness from what is specified. Other types of defects cause problems in only isolated cases.

Manufacturers of laminate who certify their material to meet MIL-P-13949 requirements must perform extensive testing to support their claim. In most cases, simply having the laminate supplier provide his lot testing results will suffice. Sometimes the laminate manufacturer performs more tests than are required and does not like to publish all of the results. In such cases, the printed circuit manufacturer can supply the laminate manufacturer with a list of properties for which test results are absolutely needed per MIL-P-5110D (Figure 7-2 contains such a listing). For each laminate type, all the laminate manufacturer need do is fill in the blanks and forward this information to the printed circuit manufacturer. The quality or manufacturing engineer need only review these forms as received for each lot of material. Note that incoming laminate inspection must be performed regardless of the documentation supplied by the laminate manufacturer.

Drill Program First Article Inspection.   When a job is programmed for drilling, it is necessary to drill one panel and submit it for inspection. The drill program cannot be released to manufacturing until the program has been validated. All second drilled, screening, and tooling (multilayer) holes must be drilled into this first article. The drill program first-article inspection consists of at least the following steps:

1. Verification of hole quantity, coding, and drill bit selection. Usually starting with the fewest number of holes per code, each hole of a given code must

INCOMING LAMINATE INSPECTION

VENDOR_____ P. O. #_____ LOT# _____

PACKING LIST INFORMATION   STICKER INFORMATION
-------------------------   -------------------

1. TYPE_____   -----------------------
                                                      N/A
2. DIMENSIONS_____   -----------------------

3. THICKNESS_____   -----------------------

4. FOIL_____   -----------------------

5. COLOR_____   -----------------------

6. CERTS ENCLOSED_____   -----------------------
                                                      N/A
9. LAB REPORT ENCLOSED--------   -----------------------

10. QUANTITY_____   -----------------------

The incoming material shall be inspected as follows:
1. Material type, foil clad and core thickness on stickers shall be as
   written on the packing slip/P.O.
2. Certification to MIL-P-13949 must be present with each lot.
3. Laboratory reports confirming certification must be present with each
lot.
4. Quantity shall be as listed on packing slip.
5. One panel size piece of laminate shall be obtained from each lot:
   A. Inspect for pits, dents and scratches.
   B. Measure the cut piece on 4 edges, at least 1 inch in from the edge,
      and record the values.
   C. Etch the piece measured and mic the same edges again, and recored the
      values.
   D. Inspect the etched piece for defects.
6. Fill in measured values and inspection results, sign and date inspection
   for and submit to QA Manager with certs and laboratory test results.

   Surface Appearance_____

   Thickness Before Etch: _____   _____   _____ _____

   Thickness After Etch: _____   _____   _____ _____

   Laminate Appearance After Etch_____

   Uneven Clad Side Identified_____

   Grain Direction Identified_____

Inspector_____ Date_____

Fig. 7-1 Incoming laminate inspection form.

be identified on the drilled panel. Also, the drill bit must reflect whether the
hole is plated or not. If the hole quantities do not agree with the blueprint,
the planning engineer must be notified.

2. The person performing the first-article inspection must review the drill

LABORATORY TEST RESULTS FOR LAMINATE

ITEM	MIL-P-13949 PARAGRAPH	VALUE SHOULD BE	VALUE MEASURED
BASE MATERIAL	1.2.1.1.1		
FOIL WEIGHT	1.2.1.1.3		
PITS/DENTS (GRADE A)	3.7.1		
THICKNESS (CLASS I)	3.5.2.1		
BOW AND TWIST (GRADE A)	3.7.3		
WRINKLES	3.7.1.2		
SCRATCHES	3.7.1.4		
APPEARANCE AFTER ETCHING	3.7.1.5		
PLASTIC SURFACE, 1 SIDED	3.7.2		
APPEARANCE: THERMAL STRESS	3.7.4		
PEEL STRENGTH	3.7.5		
DIMENSIONAL STABILITY	3.7.7		

The above test results and requirements prove that this laminate meets all requirements of MIL-P-13949 for the listed classes.

Signed_____ Date_____

Title_____

Fig. 7-2 Laboratory test results for laminate form.

data card located inside the box which stores the drill tape. It is not enough to have the drilled first-article panel. Nonplated holes which are to be second drilled must be listed for second drilling on the drill data card. Holes which are to be primary drilled and then plugged must be listed for primary drilling and identified for plugging on the traveler and blueprint. Nonplated holes which are to be primary drilled and then tented with photoresist must be called out for tenting on the artwork inspection report. Also, if a hole code is to be tented, the traveler must call out for photoresist imaging.

3. Multilayer panels must have the correct number, size, and location for all of the lamination tooling holes.

4. Plated slots must be present.

5. The number of images per panel (number up) must agree with the traveler, as must the panel size.

6. A diazo copy of the artwork is laid against the drilled panel to check for registration and missing or extra holes, that is, pads where there are no holes and holes where there are no pads.

7. Panel layout
   a. The artwork is also used to verify correct part spacing and the presence of at least a .500- to 1.0-inch border for plating.
   b. The parts should have been laid out for the gold-plated contact fingers to be on the outer edges. This is done to avoid shearing the panel needlessly. If a panel does have to be sheared for gold contact plating, there must be at least .500 inch separating the images where the shear cut will be made. There must also be a full set of screening holes for each half-panel.
   c. Multi-up panels are usually flip-flopped on the layout unless there is a specific restriction, such as a single-sided circuit or a multilayer with unbalanced dielectric spacing or foil construction.

Multilayer Processing.   First articles must be obtained and inspected for each of the following operations:

1. Inner layer imaging. The first-article inspector has the opportunity to verify the following:
   a. The need for touchup on the artwork and the imaged layer. Areas of poor image quality typically are circled with chalk and looked for on the artwork to prevent repetition of errors.
   b. The trace width must be correct within the processing guidelines of the shop and the artwork inspection report.
   c. The foil cladding and dielectric thickness of the layer must be correct as specified on the traveler.
   d. The photoresist step tablet must display the proper exposure (step held or copper step).

2. Etching—line width, line quality, and completeness of foil removal are checked.

3. Black oxided layers are checked for uniformity and quality of deposit. If the oxide can be brushed off with a paper tissue, the deposit is too thick and powdery. The oxide should be removed and the layers reprocessed (for a shorter time).

4. Lamination: All panels are checked for:
   a. Excessive image transfer (inner layer images being transferred through the outer layers as ridges of copper).
   b. Pits and dents due to unclean lamination plates.
   c. Laminated thickness being within the specified range.

A panel from the first load is etched completely. At least the following items must be checked:

1. Registration: determined visually and with the aid of a $60\times$ microscope.

2. The presence of resin dryness, air entrapment, or delamination.

3. Layer sequencing. The layer sequence numbers must be completely visible and read correctly from left to right when the panel is viewed by looking through layer 1.

4. Dielectric spacing should be verified on difficult or important jobs, especially those containing 2-oz foil on one or more layers. It is necessary to trim off and microsection a coupon if the job has specific dielectric tolerances which must be held.

First Article Production Drilling.   The drilling operator must use a micrometer to verify drill sizes. The operator will also use a drilled mylar copy of the programming first article to verify registration, shallow drilled holes, and broken drill bits. When setting up for drilling multilayer panels, an etched panel is used to set the zero point and to verify the accuracy of registration on each drill spindle. The multilayer drill first article must always be approved by the drilling supervisor.

Electroless Copper.   A panel from the first lot run is taken to the inspection department. Here the holes and edges are checked for coverage. The deposit is tested for peeling electroless copper. Poor coverage and a peeling deposit must be reported to the plating supervisor. If the panel is a multilayer, it may be necessary to check for etchback (if required). At the very least, the inspector must ask the electroless copper plater if the panel was processed through desmear or etchback. Desmear is absolutely required for all multilayer panels. Etchback is required only when specified by the customer. See Chapter 22 of the *Handbook of Printed Circuit Manufacturing*, for a discussion of desmear and Etchback.

Imaging.   The first images from dry film or screen printing must be submitted to either the inspection or touchup department for approval. The first dry film imaged panels must also contain a step tablet reading which denotes the correct degree of exposure (See "Dry Film Imaging" in the *Handbook of Printed Circuit Manufacturing.*) It is good practice to use a 60× microscope. At least the following items must be verified:

1. The trace width should be within .001 inch of the artwork.

2. The minimum air gap (conductor spacing) must meet the same requirement.

3. The minimum annular ring must meet the specification.

4. Registration. The artwork registration to the drilled hole pattern must be within .002 or .003 inch (whatever is accepted for shop processing tolerance) of the registration noted on the artwork inspection report. If .000 layer-to-layer misregistration is noted, the image on the panel should be within the shop's tolerance. If .002 inch of misregistration is noted on the artwork inspection report, the image on the panel may be off as much as .002 inch plus the shop tolerance.

5. Resist quality—dry film. The edge definition must be solid, with no lifted edges or undeveloped resist.

6. Resist quality—screened on ink. The edges must be free of shadowing and skips.

7. Registration must be at least as good as the artwork inspection report allows. The annular ring must be at least what has been measured on the artwork inspection report.

Plating.   First articles are required at several operations in the plating department. Some of these operations may be performed by the plating supervisor, others by the inspection department. Unacceptable conditions must always be reported to the plating supervisor or another supervisor.

1. Hole plugging. Prior to plating, if hole plugging is required, one panel should have all of the required holes plugged. The blueprint is always required for locating holes to be plugged. Once the holes have been plugged, this first-article panel serves as an example for the rest of the panels. The plating supervisor usually inspects this operation.

2. First-article plating (also strip, etch, and reflow). This is performed to verify that the proper amounts and types of metals have been plated and that the

holes are the correct size. Often a through-hole measuring device is used to measure electrically the copper plated in the holes. These devices work only after the copper foil has been etched. For this reason, it is common to take a panel from the first plating load to strip the resist, etch the foil, and fuse (reflow) the tin-lead. Fusing the tin-lead allows accurate measurement of hole size and trace width. The inspection of a plating frist article proceeds follows:

a. Hole sizes are measured for each hole code. Nonplated holes must also be checked for the presence of plating. Plating will enter a nonplated hole if resist tents have broken, if plugs have fallen out, or if holes were incorrectly plugged.

b. Through-hole copper thickness is measured using electrical equipment built for that purpose.

c. Tin-lead or other metals can also be checked by cutting and performing a microsection. Through-hole quality, inner layer connections, metal thicknesses, and dielectric spacing can also be determined accurately at this time. Tin-lead thickness can also be determined to be acceptable or unacceptable by its appearance. Rounded, smooth, shiny solder generally indicates acceptable thickness. Falt, dull, dewetted solder generally indicates inadequate thickness.

d. Holes must always be checked visually. This will reveal voids and other conditions which can be corrected prior to running the entire production lot.

   (1) A black coating inside the holes may indicate the need to change the hot oil used for reflow. Plugged holes or holes with a black ring also indicate the need for a change of the fusing oil.

   (2) A white deposit inside the holes may indicate that the solder conditioner on the etching machine is spent and needs to be changed.

3. Trace width must be checked with a $60\times$ microscope. Jobs imaged with dry film photoresist often lose .001 to .002 inch of trace width per ounce of copper foil. A loss of more than this amount may indicate overetching or a dry film exposure problem. If the trace width is identical to the artwork on the unplated panels, or on the plated but unetched panels, reduced traces are most likely the result of overetching.

4. Gold-plated contact fingers are checked for a number of conditions, all of which will cause scrap if not detected early and corrected.

   a. Nickel and gold thicknesses are measured on electrical and beta-back-scatter equipment designed for this purpose.

   b. Cellophane tape is used to check for metal peeling. If slivers of nickel/gold from the overhanging edges of the contact fingers are present on the tape, this is acceptable. No other metal is allowed to come off on the tape.

c. The demarcation line between gold and solder should be clean. A thin (.030 inch or less) black or tarnished line is acceptable but does not indicate optimum processing. If repressing of the plater's tape can eliminate an ugly demarcation line, the panels should be repressed.

d. Black holes above the plater's tape line, or beneath the tape at the contact fingers, are unacceptable. Excessive fuming of solder stripper and sloppy processing cause black holes above the tape line. Allowing more space between panels in the solder stripping tray often eliminates this condition. Excessive time in the solder stripper and poor tape pressing cause black holes beneath the tape at the contact fingers. Tape should be repressed if more than a shift (8 hours) has elapsed between pressing and solder stripping.

e. Plater's tape residue above the contact fingers is unacceptable, and platers must be notified of this fact.

Solder Mask.   Solder mask first articles must be checked for a variety of conditions. It is usually a good idea to follow up solder mask first articles by roving inspection. The first few panels may look acceptable, only to be followed by excessive mask bleeding, smearing, and registration problems afterward. Items to be checked include:

1. Type of solder mask and color of mask.
2. Number of sides on which it is required—if it is required at all.
3. Exposed traces. Traces may be exposed due to excessive clearance on artwork, wrong artwork used (round versus oval clearances), or misregistration.
4. Mask on pads and/or surface mounting pads. Mask on pads may also result in an electrical testing problem. Mask on surface mounting pads is almost always unacceptable.
5. Clearance zones provided in the mask artwork per the blueprint.
6. Solder mask in holes, both plated and nonplated.
7. Both sides must be checked by first-article inspection, one at a time.
8. The cure can be checked, usually at the completion of a job, by rubbing Q-tips dipped in methylene chloride on the mask. If the green mask readily comes off, the cure is questionable.
9. Adhesion must be checked by tape testing. Usually, if the mask sticks when tape is applied to small traces, the adhesion is adequate. Solder mask will almost always peel to some extent when tape is applied to large ground areas. There is a destructive crosshatch cutting/tape test described in IPC literature which is not recommended by this author.

10. Excessive skips (dryness between traces) is generally not acceptable. Slight skipping in isolated areas, where trace edges are clearly covered, should be acceptable.
11. Excessive bubbles in the mask indicate that the squeegie stroke was too rapid. This condition should not be tolerated. While it may not be cause for rejection from the customer, it should be rejected at the first-article stage. Bubbles in the solder mask indicate poor screen printing technique.
12. Solder mask on the tips is generally acceptable when it covers only the top rounded portion of the tips. Splashes of mask and mask along the edges of the contact fingers are not acceptable.

Legend (Epoxy Silk Screen).   Legend first articles are just as critical as solder mask first articles. Items to check for include:

1. Legend applied to the correct side(s).
2. Legend registered correctly. Registration and application to the correct sides are often difficult to determine. The planning engineer must be aware of this during the artwork inspection process. Special steps may be necessary to make registration reliable, such as the addition of targets to the artwork and blueprints.
3. Readability of all characters.
4. Peeling legend must be checked using the tape test.
5. Legend ink in holes or on pads and surface mount pads. Legend on surface mount pads may cause rejection. Legend on through-hole pads may cause an electrical testing problem.
6. Color of legend ink.
7. Bubbles in the legend. These look bad and should not be tolerated at the first-article stage. Too rapid squeegie stroke causes this condition.

Second Drill and Fabrication.   The act of drilling holes after plating is called ''second drilling.'' It is one of the methods used to produce nonplated holes. Second drilling is most often performed after solder masking, just prior to fabrication (depanalization). Second-drilled holes may be referenced from a datum or other point denoted on the blueprint.

Because the laminate has been through a number of operations, it may have undergone dimensional changes. Also, since the panels have been removed from the drilling machine after primary drilling, it is possible to place the panels back on the drill incorrectly and to misset the zero, use the wrong drill tape, or make any number of other mistakes. It is good practice to obtain a first-article inspection before drilling all of the panels in a given job.

Fabrication is the act of routing the final board outline on a printed circuit. There is no excuse for routing an entire job, only to find out later that the router bit compensation was off or the router tape contained an error. The second drill and fabrication first articles may be the same. All dimensions on a blueprint should be written on a piece of paper (or a form designed for this purpose) and the actual measured dimensions from the first article written next to the specified value.

Final Inspection/Preshipping Verification. Before any job leaves the final inspection department, a first-article form must be completed for each part number and manufacturing lot (Figure 7.3 shows an example of this type of form). Such a form will cover about 95% of the most common types of errors. This form provides a written history of each lot and may come in very handy later on.

Summary. First articles, when mandatory at each operation, can prevent quality and manufacturing problems from affecting the entire lot. The first article should usually be done by a person outside the department. This outsider can often spot problems which go unseen by persons who are looking at one operation all the time. Widespread use of first articles also tends to keep operators and supervisors alert and honest. Departmental personnel will take the time to shoot a step tablet for dry film or tear down a bad screen stencil if they know that their first article is likely to be rejected by an inspector. There are too many benefits to be derived from first articles to allow any department to proceed without them.

Electrical Testing and Automated Optical Inspection (AOI). These techniques are widely used to assure quality. Typically, AOI is used for inner layer inspection and electrical testing is used to test the completed board.

AOI of Inner Layers. With AOI, the printed circuit manufacturer can achieve optimum final testing yields on even the most complicated multilayer printed circuits. When the inner layers are known to contain no opens or shorted circuits, electrical test failures can only be due to outer layer problems, except for plated-through hole voids, delamination, or misdrilling. AOI will catch very narrow cuts in traces, nicks, and other items likely to be missed by human inspection. AOI is fast (less than a minute per layer) and will catch some problems which electrical testing of inner layers will miss, such as extraneous copper which will cause a short circuit after lamination and drilling.

The chief drawback to AOI is that there is a certain amount of operator dependency. The care taken in setting up the machine can vary considerably from person to person, as can the results obtained. Also, AOI will identify conditions which are not really faults, such as water spots and oxidation. It is common for layers to be inspected and tagged with many more ''defects'' than really exist.

PRE-SHIP INSPECTION REPORT

CUSTOMER_____ PART NO. _____ REV_____

JOB NO. _____ DRAWING NO. _____ REV_____

DUE DATE_____ ARTWORK NO. _____ REV_____
==

GOLD: THICKNESS   S/B_____ IS_____   TAPE TEST_____

LAMINATE: TYPE   S/B_____ IS_____   MIN. TRACE: S/B_____ IS_____

        : THICKNESS   S/B_____ IS_____   MIN. AIRGAP: S/B_____ IS_____

SOLDERMASK: SIDES   S/B_____ IS_____   MIN. ANN. RING S/B_____ IS_____

         : COLOR   S/B_____ IS_____   LEGEND: SIDES S/B_____ IS_____

         : RELIEVED PER DRAWING?_____   : COLOR S/B_____ IS_____

         : TRACES EXPOSED?_____   : REGISTER CORRECTLY?_____

         : MASK ON PADS?_____   : READABLE?_____

         : TAPE TEST_____   : TAPE TEST_____

       FABRICATION DIMENSIONS   HOLE SIZES
    S/B   IS   S/B   IS

APPROVED FOR SHIPPING _____ DATE _____

SPECIAL INFORMATION:

Fig. 7-3 Preship inspection report.

Therefore, the people responsible for verifying and reworking must develop some degree of judgment. With practice, experienced personnel become proficient in making AOI equipment serve the needs of the printed circuit manufacturer.

The customer pays for testing of the finished printed circuit board. How the printed circuit manufacturer builds a good board is not the customer's concern as long as the board is capable of passing all tests—electrical, metallurgical, and otherwise. AOI of inner layers is something the printed circuit manufacturer does for his own benefit. If this service must be contracted out, it can be costly. The manufacturing engineer must decide when a given job should be sent for AOI of inner layers.

Electrical Testing of Inner Layers. Electrical testing of inner layers works best when all of the holes, including vias, are on .100-inch centers. The grid in universal test equipment is also on .100-inch centers. When testing inner layer pads on .100-inch centers, the pins and pads align easily. Some problems arise in attempting to test electrically inner layers which contain pads, especially via hole pads, not on .100-inch centers. The fixture pins must be positioned at an angle. When this happens, pins touch each other and misalign to the pads. The result is intermittent false readings. It may even be impossible to program or test. For this reason, AOI is unsurpassed for inner layer quality assurance.

There is another consideration when deciding to test inner layers electrically. Since testing is performed by placing the layer on a bed of pins, a second fixture must be built for testing patterns on the other side of the layer. This second fixture is the mirror image of the first.

Electrical Testing of Finished Printed Circuit Boards. There are two main types of electrical test equipment used by the printed circuit manufacturer: universal and dedicated. Universal electrical testers have made it possible to test economically almost every job being built in the shop—even the smallest prototype quantities. All that is needed to build a fixture are one or two pieces of drilled plexiglass (actually, polycarbonate is used because it has greater rigidity) and a quantity of reusable pins. Fixtures can be built in a matter of hours. The only drawback of universal testers is speed. Compared to dedicated equipment, the universal tester is slow.

Dedicated electrical testers have been around much longer than universal testers. They require that pins be wire wrapped into the test fixture. This is a time-consuming operation. The pins must be placed by hand and then wire wrapped. A universal test fixture can have the pins inserted by a shaker table in a few hours, compared to several days for the dedicated fixture. Testing speed is limited mostly by how rapidly the operator can lift a lid, place a printed circuit on it, and close it. For this reason, when large quantities are involved, the more costly dedicated test fixture is built.

Appendix A

MIL-STD-275E: Printed Wiring for Electronic Equipment

This is an important and commonly referenced industry standard. Many companies base their own internal design standards on this document. All incoming artwork and blueprints should be inspected against the criteria established here. The document is divided into six sections plus an appendix. The sections are described below. During the discussion of each section, references are made to various tables and paragraphs. If the specified table or paragraph is not reproduced in the discussion, it is found in the military standard itself.

SECTION 1. SCOPE

The scope and purpose of this standard are explained by paragraph 1.1 and 1.2 of MIL-STD-275E. This standard establishes design requirements governing rigid printed wiring boards. It establishes three classifications of boards:

Class 1—single-sided boards
Class 2—double-sided boards
Class 3—multilayer boards

Also covered are design standards for printed wiring assemblies (using the above types of boards) and design considerations for mounting parts and assemblies thereon. This standard assumes that all of the above boards will be conformally coated, in accordance with MIL-I-46058.

SECTION 2. REFERENCED DOCUMENTS

There is a tabulation of documents published by the federal government and branches of the military, American National Standards Institute (ANSI), American Society for Testing and Materials (ASTM), Institute of Electrical and Electronics Engineers (IEEE), and the Institute for Interconnecting and Packaging

159

Electronic Circuits (IPC). These documents are referenced throughout MIL-STD-275E and other specifications and standards. They form part of the text. Where they conflict with this standard, the provisions of MIL-STD-275E take precedence.

The documents referenced cover such items as materials, drawings, film preparation, plating, testing, definitions, packaging, marking, and other important considerations. Inclusion of references to these documents allows for greater clarity and brevity in this standard or any other standard which makes reference to them.

SECTION 3. DEFINITIONS

This section simply states that "The terms and definitions used herein shall be in accordance with this standard and IPC-T-50."

SECTION 4. GENERAL REQUIREMENTS

This section establishes a number of requirements which all planners, designers, and quality assurance people must be aware of:

1. Quality conformance test coupons must be included in the design and layout of the printed circuit board. The requirements of these coupons are listed in Section 5.9 of this standard.
2. The contents of the entire set of drawings for the printed circuit boards and assemblies are called out. Figure A-1 shows what must be covered by the set of drawings and indicates their relationships.

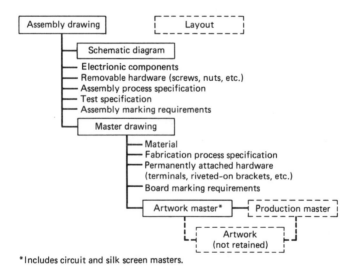

*Includes circuit and silk screen masters.

Fig. A-1 Block diagram depicting typical printed-wiring drawing relationships.

3. Component reference designations must meet the requirements of IEEE-STD-200.
4. Procedures to be followed by designers who are requesting deviations from the requirements of this standard are listed.
5. Items which must be covered by the master drawing are listed in paragraph 4.3 (See Table A-1). There is a list of 22 items, which are also important to the printed circuit manufacturer, since he/she cannot properly plan for manufacturing without knowing what these requirements are.
6. Hole location tolerances for various types of holes are discussed, such as plated-through holes, tooling holes, mounting holes, windows, access holes, via holes, and component holes.
7. Processing allowances are discussed and defined. Table A-2 is a composite of design features and their manufacturability. Manufacturability is expressed as follows:
   a. Preferred—easiest to conform to, most cost effective.
   b. Standard—greater difficulty of conformance.
   c. Reduced producibility—not likely to be attained without greater effort, care, and/or expense.
   Table A-2 is such an import design summary that it is reproduced below in full. All manufacturing planners, designers, and quality assurance people should be familiar with its content and refer to it as needed. It is a mistake to ignore the contents of this table.
8. The use of a datum point is established. Two mutually perpendicular lines are drawn through a single hole. All dimensions for holes, slots, hardware, board outlines, etc. are referenced from these datum lines. The use of these datum lines makes it possible to establish acceptable tolerances.
9. Assembly drawing requirements are listed in paragraph 4.4. This book is primarily concerned with planning for printed circuit manufacturing. However, there is frequently some amount of assembly which a board manufacturer will perform. For this reason, paragraph 4.4 is reproduced (see Table A-3).
10. Production masters, film for each layer needed to build the printed circuit, shall be prepared on .0075-inch, dimensionally stable film according to the requirements of MIL-D-8510, type II, L-F-film.

SECTION 5. CONDUCTOR PATTERN

This section covers the design requirements for all conductive patterns, dielectric spacing, material thickness tolerance, and other physical features. This section is important; all planning, design, and quality assurance people must have an understanding of the information covered here. Paragraph 5.1.1 establishes

Table A-1   Required Information for the Master Drawing

The master drawing shall be prepared in accordance with DOD-STD-100; shall include all appropriate detail board requirements (see Section 5), and the following:

a. The type, size, and shape of the printed wiring board.

b. The size, location, and tolerance of all holes therein.

c. Etchback allowances, when required or permitted.

d. Location of traceability marking.

e. Dielectric separation between layers.

f. Shape and arrangement of both conductors and nonconductor patterns defined on each layer of the printed wiring board. Copies of the production masters or copies of the artwork may be used to define these patterns.

g. Separate views of each conductor layer.

h. Any and all pattern features not controlled by the hole sizes and locations shall be dimensioned either specifically or by reference to the grid system (see n).

i. Processing allowances that were used in the design of the printed wiring board (see 5.1.1, 5.1.4, 5.2.1.2, 5.2.2, and 5.2.2.6).

j. All notes either included on the first sheet(s) of the master drawing or by specifying the location of the notes on the first sheet.

k. Conductor layers numbered consecutively, starting with the component side as layer 1. If there are no conductors or lands on the component side, the next layer shall be layer 1. For assemblies with components on both sides, the most densely populated side shall be layer 1.

l. Identification marking (see 5.8).

m. Size, shape, and location of reference designation and legend markings, if required (see h).

n. A modular grid system to identify all holes, test points, lands, and overall board dimensions with modular units of length of 0.100, 0.050, 0.025, or other multiples of 0.005 inch in that order of preference. For designs where the majority component locations are metric based (SI), the basic modular units of length shall be 2.0, 1.0, 0.5, or other multiples of 0.1 mm in that order of preference. The grid system shall be applied in the X and Y axes of the Cartesian coordinates. The grid shall not be reproduced on the master drawing; but may be indicated using grid scales or X, Y control dimensions.

o. Dimensions for critical pattern features which may effect circuit performance because of distributed inductance or capacitance effects within the tolerance required for circuit performance.

p. All terms used on the master drawing shall be in conformance with the definitions of ANSI/IPC-T-50 or ANSI Y14.5 (see 3.1 and 2.2).

q. Deviations to this standard (see 4.2.2).

r. Minimum line width and spacing of the finished printed-wiring board.

s. Maximum rated voltage (maximum voltage between the two nonconnected adjacent conductors with the greatest potential difference) for type 3 boards only.

t. Plating and coating material(s) and thickness(es).

u. Identification of test points required by the design (see 5.1.8).

v. Applicable fabrication specification with date(s), revision letter, and amendment number.

an important concept: the need to compensate artwork for processing tolerances so that the finished product may come as close as possible to nominal design requirements. Table A-2 summarizes much of what is discussed in this section.

Table A-2 Composite Board Design Guidance

	Preferred	Standard	Reduced Producibility
Number of conductor layers (maximum[1])	6	12	20
Thickness of total board (maximum) (inch)	.100 (2.54)	.150 (3.81)	.200 (5.08)
Board thickness tolerance	±10% of above nominal or 0.007 (.18), whichever is greater		
Thickness of dielectric (minimum)	.008 (.20)	.006 (.15)	.004 (.10)
Minimum conductor width (or Figure 4 value, whichever is greater)			
Internal	.015 (.38)	.010 (.25)	.004 (.10)
External	.020 (.51)	.015 (.38)	.004 (.10)
Conductor width tolerance			
Unplated 2 oz/ft^2	+.004 (.10)	+.002 (.05)	+.001 (.025)
	−.006 (.15)	−.005 (.13)	−.003 (.08)
Unplated 1 oz/ft^2	+.002 (.05)	+.001 (.025)	+.001 (.025)
	−.003 (.08)	−.002 (.05)	−.001 (.025)
Protective plated (metallic etch resist over	+.008 (.20)	+.004 (.10)	+.002 (.05)
2 oz/ft^2 copper)	−.006 (.15)	−.004 (.10)	−.002 (.05)
Minimum conductor spacing (or Table I, whichever is greater)	.020 (.51)	.010 (25)	.005 (0.13)
Annular ring plated-through hole (minimum)			
Internal	.008 (.20)	.005 (.13)	.002 (.05)
External	.010 (.25)	.008 (.20)	.005 (.13)^2
Feature location tolerance (master pattern, material movement, and registration (rtp))			
Longest board dimension 12 inches or less	.008 (.20)	.007 (.18)	.006 (.15)
Longest board dimension over 12 inches	.010 (.25)	.009 (.23)	.008 (.20)
Master pattern accuracy (rtp)			
Longest board dimension 12 inches or less	.004 (.10)	.003 (.08)	.002 (.05)
Longest board dimension over 12 inches	.005 (.13)	.004 (.10)	.003 (.08)
Feature size tolerance	±.003 (.08)	±.002 (.05)	±.001 (.025)
Board thickness to plated hole diameter (maximum)	3:1	4:1	5:1
Hole location tolerance (rtp)			
Longest board dimension 12 inches or less	.005 (.13)	.003 (.08)	.002 (.05)^3
Longest board dimension over 12 inches	.007 (.18)	.005 (.13)	.003 (.08)^3
Unplated hole diameter tolerance (unilateral)			
Up to 0.032 (0.81)	.004 (.10)	.003 (.08)	.002 (.05)
0.033 (0.84)–0.063 (1.61)	.006 (.15)	.004 (.10)	.002 (.05)
0.064 (1.63)–0.188 (4.77)	.008 (.20)	.006 (.15)	.004 (.10)
Plated hole diameter tolerance (unilateral) for minimum hole diameter maximum board thickness ratios greater than 1:4 add 0.004 (0.01)			
.015 (.38)–0.030 (.76)	.008 (.20)	.005 (.13)	.004 (.10)
.031 (.79)–0.061 (1.56)	.010 (.25)	.006 (.15)	.004 (.10)
.062 (1.59)–0.186 (4.75)	.012 (.31)	.008 (.20)	.006 (.15)
Conduction to edge of board (minimum)			
Internal layer	.100 (2.54)	.050 (1.27)	.025 (.64)
External layer	.100 (2.54)	.100 (2.54)	.100 (2.54)

[1] The number of conductor layers should be the optimum for the required board function and good producibility.

[2] See 5.2.3.

[3] To be used only in extreme situations warranted by the application.

NOTE: Unless otherwise specified, all dimensions and tolerances are in inches; data in parentheses () is expressed in millimeters.

Other tables and figures referred to herein are found in MIL-STD-275E. The reader is advised to obtain copies of this, and other standards, listed in the appendixes.

Table A-3   Required Information for the Printed Wire Assembly Drawing

The printed wiring assembly drawing shall cover printed wiring on which separately manufactured parts have been added. The printed wiring assembly drawing shall be in accordance with DOD-STD-100 and should include at least the following:

a. Parts and material list.
b. Component mounting and installation requirements.
c. Cleanliness requirements per MIL-P-28809.
d. Location and identification of materials or components (or both).
e. Component orientation and polarity.
f. Applicable ordering data from MIL-P-28809.
g. Structural details when required for support and rigidity.
h. Electrical test requirements.
i. Marking requirements.
j. Electrostatic discharge protection requirements.
k. Special solder plug requirements.
l. Eyelets and terminals.
m. Lead forming requirements.
n. Type of conformal coating and masking.
o. Solder mask.
p. Traceability.

The printed wiring assembly drawing shall include the definition of any conditions considered in the design where the manufacturing variation between the end product and the assembly configuration plays a role in the producibility or performance of the printed wiring assemblies.

Design Features

1. Line widths on master artwork must be compensated for processing allowances to meet or maintain conductor width on the master drawing.

2. Conductors which change direction with less than 90 degree angles shall have the external corner of the bend rounded.

3. Conductor lengths should be held to a minimum. Conductors running along the X axis, the Y axis, and at 45 degrees to these axes are preferred to facilitate computer-aided designing.

4. Conductor spacings shall be as large as allowable. Minimum spacing shall be kept in accordance with Table A-1 of MIL-STD-275E. This table references conductor spacing for internal and external layers as a function of voltage.

5. Minimum spacing between conductors and the board edges shall be as listed in Table A-I of this specification plus .015 inch.

6. Large conductive areas and the problems they cause are discussed. The layout of large metal areas is specified, along with the use of nonfunctional metal areas to balance the construction.

7. Interfacial connections are to be made by the plated-through hole, not by eyelets of other types of hardware.

8. Holes are to be designed to facilitate solder wicking into holes around component leads and to provide solder plugs of holes without leads. Paragraph 5.1.7.1 has an extensive discussion of the requirements for forming solder fillets and solder plugs or plugging the holes with solder mask.

9. The presence of test points, when required by the design, is discussed. These shall be plated-through holes.

10. The minimum annular ring (land area around a plated or nonplated hole) is discussed, as well as the method used to calculate it (paragraph 5.2.1.2).

11. Minimum land area for surface mounted components are discussed.

12. Eyelets are specifically forbidden in new designs without approval of the federal government (paragraph 5.3.3).

13. Drilling directly into internal power and ground planes is forbidden, without some kind of thermal relief (see Figure 14 in the Appendix, not reproduced here).

14. All hole and land areas shall be located on a grid intersection specified on the master drawing. It is permissible to have holes in a group which are off this master grid pattern, provided at least one hole in that group is on the grid and serves as a datum for the other holes in the grouping.

15. Specifications for eyelets and other forms of hardware are discussed (see paragraphs 5.4 to 5.4.3).

16. All metal-clad laminates shall be per MIL-P-13949.

17. Minimum laminate thickness is discussed in paragraph 5.6.1.1.

18. Prepreg bonding materials shall be per MIL-P-13949. GE and GF prepreg (epoxy/fiberglass) shall not be used with GI (polyimide) prepreg.

19. Finished conductors must have at least 1 oz of thickness (.0012 inch) and must be made from at least $\frac{1}{2}$ oz copper foil plated to at least 1 oz finish.

20. Plating requirements (paragraph 5.6.4)

   a. All external conductive patterns shall be covered with solder, unless covered by solder mask, a heat sink, or other approved plating.

   b. When other metal plating is approved, no copper shall be exposed as the interface of the two metals.

   c. No other metals shall be plated over tin-lead or tin.

   d. Unless otherwise specified on the master drawing, through-hole and surface metal thicknesses shall be per paragraphs 5.6.4 through 5.6.4.6.

21. Solder mask coating shall be per class 3 of IPC-SM-840. It shall not be used unless specified on the master drawing. The use of chemical treatments to improve solder mask adhesion is specifically allowed per paragraph 20.5 of the Appendix.

22. Board thickness shall include plated metals and solder mask coating.

23. Multilayer printed circuits shall have a minimum dielectric spacing of .0035

inch between conductive layers. There must be at least two sheets of pre-preg.

24. Maximum bow and twist shall be 1.5%.

SECTION 6. DETAIL PART MOUNTING REQUIREMENTS

This section presents in-depth descriptions of design parameters which affect assembly. The information here is of great importance to printed circuit designers and to quality assurance people involved with design approval and assembly. It will not be covered in depth here because this book is primarily concerned with manufacturing the bare printed circuit board once it has been designed and artwork generated.

SECTION 10. APPENDIX

The purpose of the Appendix is to provide further guidance to the designer of printed wiring boards. Planners and quality assurance people should be familiar with much of the information presented here. The Appendix contains the following:

1. The Table A-2 composite design summary (reproduced in this chapter as well).
2. Design and layout of quality conformace test coupons. These coupons are required to be present on the working film of the printed circuit board manufacturer.
3. A block diagram depicting the relationship of all drawings for printed circuit boards.
4. A diagram of how a grid pattern is used to define pattern requirements.
5. Graphs to be used for determining conductor widths and copper foil thicknesses as a function of amperes and the resulting temperature of the conductor.
6. Land pattern requirements for surface-mounted chip carriers.
7. Pad shapes and minimum annular ring measuring requirements.
8. Where to measure minimum dielectric spacing between conductive layers.
9. Where to locate quality conformance coupons on the printed circuit panel as a function of the number of boards on the panel.

Appendix B

IPC-D-300G: Printed Board Dimensions and Tolerances

This is an invaluable document for anyone who needs to understand some of the most basic information presented on blueprints: dimensions and tolerances. Those who must understand the information presented here include designers, quality assurance people, and manufacturing planners. Measurements are presented in metric units, together with U.S. conversions to inches.

The information presented in this standard is based on industry capability, which, of course, must not be disregarded. This document contains five section, including the Appendix: (1) Scope, (2) Applicable Documents, (3) Requirements, (4) Summarization, (50) Appendix.

SECTION 1.0 SCOPE

The purpose of this section is expressed in paragraph 1.1 as follows: "The purpose of this standard is to establish rules, principles, and methods of dimensioning and tolerancing used to define the end product requirements of a printed board on a master drawing." It establishes three classes (A, B, and C) of progressively more difficult requirements. These classes should not be confused with classes 1, 2, and 3 of end item use, which are listed in other IPC standards and specifications. Selection of dimensional classes A, B, and C should be determined by the minimum need for precision, whereas selection of classes 1, 2, and 3 should be based on the purpose of the end product: consumer, general industrial, or high reliability.

IPC-D-300G provides for two types of dimensioning systems:

Type 1—Nominal dimensioning (tolerances apply to dimensions).

Type 2—Basic dimensioning (tolerances are expressed as an allowable variation from the basic dimensions specified).

Paragraph 1.5 lists numerous terms and definitions which have precise technical meaning.

SECTION 2.0 APPLICABLE DOCUMENTS

This section lists other documents which contain information necessary to understand fully the scope of IPC-D-300G, since IPC-D-300G is also referenced in other documents.

SECTION 3.0 REQUIREMENTS

1. The main features that are typically dimensioned and toleranced are listed below, from paragraph 3.1. These are:

   a. Minimum conductor width
   b. Minimum conductor spacing
   c. Minimum annular ring
   d. Lands
   e. Plated-through holes
   f. Nonplated-through holes
   g. Printed board length, width, and thickness
   h. Connectors

2. Paragraph 3.1.1 discusses the use of data and provides diagrams of several examples.

3. All holes, lands, and features of printed circuits shall be dimensioned by the use of a grid system, except when necessary for mating with parts that are not on a grid system. Preferred grids are .100, .050, and .025 inch.

4. The annular ring is defined as the minimum distance from the edge of a functional land to the edge of the drilled hole (single-sided boards and inner layer of multilayer boards) or to the inner edge of a plated-through hole for all others. Table B-1 lists the minimum requirement of the annular ring for classes A, B, and C.

Table B-1   Annual Rings (Minimum)

Annular Ring	Class A	Class B	Class C
Internal supported	.15 [.006]	.05 [.002]	.03 [.001]
External supported	.25 [.010]	.15 [.006]	.05 [.002]
External unsupported	.40 [.016]	.25 [.010]	.15 [.006]

5. Minimum land size around a hole is determined by considering the following:
   A. Maximum diameter of the drilled hole.
   B. Minimum annular ring requirement.
   C. Maximum allowance for etchback, when required.
   D. A standard manufacturing allowance, which must take into account tooling and processing variations. Table B-2 lists standard tolerances which can be applied to each class, depending on the size of the board.
   This is summarized by the formula below and discussed in much greater detail in the appendix.

$$\text{Minimum land} = A + 2B + D + 2C \text{ (if required)}$$

6. Bow and twist is discussed in paragraph 3.1.4, together with information on each tolerance class (see Table B-3).

Table B-2   Standard Manufacturing Allowances

Greatest Board/Panel Dim.	Class A	Class B	Class C
Up to 300 [12.00]	.70 [.028]	.50 [.020]	.30 [.012]
More than 300 [12.00]	.85 [.034]	.60 [.024]	.40 [.016]

Table B-3A   Bow and Twist Tolerance, Paper Base and Composite Materials

Pattern	Thickness Code	Class A	Class B	Class C
1				
S	T1	No req.	2.5%	1.5%
I	T2	2.5%	2.0%	1.0%
D	T3	2.0%	1.2%	0.8%
E	T4	1.5%	0.8%	0.6%
D				
2				
S	T1	No req.	2.0%	1.0%
I	T2	2.0%	1.5%	0.8%
D	T3	1.5%	1.0%	0.6%
E	T4	1.0%	0.7%	0.7%
D				

Table B-3B   Bow and Twist Tolerance, Glass Base Material

Pattern	Thickness Code	Class A	Class B	Class C
1				
S	T1	2.5%	2.0%	1.5%
I	T2	2.0%	1.5%	1.0%
D	T3	1.5%	1.0%	0.8%
E	T4	0.8%	0.6%	0.6%
D				
2				
S	T1	2.0%	1.5%	1.5%
I	T2	1.5%	1.0%	0.9%
D	T3	1.0%	0.7%	0.6%
E	T4	0.6%	0.5%	0.5%
D				
M				
U				
L	All categories	3.0%	2.0%	1.0%
T				
I				

7. Type 1, the Nominal Dimensioning System, is discussed beginning with paragraph 3.2. This system assigns a desired (nominal) dimension to each feature of the board and then assigns a tolerance to that dimension. Typically, a board edge serves as a datum.

There is a discussion of each type of dimension and a table with the tolerance values of each. The types of dimensions listed are:

a. Board edges (see Table B-4).
b. Board connector tangs (contact finger areas).
c. Cutouts, notches, and keying slots.

   NOTE: Radii should be provided for all slots and notches.

d. Board thickness (see Table B-5).
e. Board edge chamfering.

Table B-4   Board Edge Tolerances

Class A	Class B	Class C
±.40	±.25	±.15
[±.016]	[±.010]	[±.006]

Table B-5   Board Thickness Tolerances

Thickness	Class A	Class B	Class C
T1	±.20 [±.008]	±.10 [±.004]	±.05 [±.002]
T2	±.30 [±.012]	±.20 [±.008]	±.10 [±.004]
T3	±.40 [±.016]	±.25 [±.010]	±.15 [±.006]
T4	±15% of nom.	±10% of nom.	±5% of nom.

For edgeboard connector, use classes B and C.

f. Unsupported hole diameters (see Table B-6).
g. Unsupported hole-to-lead ratio.
h. Finished diameter of the plated-through hole (see Table B-7).

NOTE: More tolerance is added over that shown in Table B-7, if the hole diameter is less than one-third to one-fourth of the board thickness.

NOTE: For any class, the larger the hole, the greater the tolerance; three size ranges are listed.

i. Supported hole-to-lead ratio.
j. Hole, feature location (see Table B-8).

NOTE: The information in Table B-8 was developed for epoxy/fiberglass. Less dimensionally stable materials may require that more tolerance be allowed.

Table B-6   Unsupported Holes

Hole Dia.	Class A	Class B	Class C
0–.8 [0-.032]	±.08 [±.003	±.05 [±.002]	±.03 [±.001]
.85–1.6 [.033-.063]	±.10 [±.004]	±.08 [±.003]	±.05 [±.002]
1.65–5.0 [.064-.188]	±.15 [±.006]	±.10 [±.004]	±.08 [±.003]
Ratio of min. hole dia. to base material thickness	2 : 3 or 66%	1 : 2or 50%	1 : 4 or 25%

Table B-7 Plated-Through Hole Diameter Tolerances

Hole Dia.	Class A	Class B	Class C
0–.8 [0–.032]	±.10 [±.004]	±.08 [±.003]	±.05 [±.002]
.81–1.6 [.033–.063]	±.15 [±.006]	±.10 [±.004]	±.05 [±.002]
.61–5.0 [.064–.195]	±.20 [±.008]	±.15 [±.006]	±.10 [±.004]

Table B-8 Hole Location Tolerances

Related Board Size	Class A	Class B	Class C
Where greatest dimension is less than 300.0 [12.00]	±.15 [±.006]	±.10 [±.004]	±.05 [±.002]
Where geratest dimension is greater than 300.0 [12.00]	±.20 [±.008]	±.15 [±.006]	±.10 [±.004]

k. Conductive pattern feature location tolerance (see Table B-9).

NOTE: Minimum annular ring requirements are a measure of conductor pattern registration.

l. Solder mask apertures feature location tolerance (see Table B-10).
m. Conductor width and spacing tolerances (see Table B-11).

NOTE: Tolerances listed in Table B-11 are for 1-oz copper conductor thickness. Allow .001 inch conductor variation for each additional ounce of conductor thickness.

NOTE: Conductor spacing requirements are the inverse of the conductor widths; apply the same tolerances.

Table B-9 Conductive Pattern Location Tolerances

Related Board Size	Class A	Class B	Class C
Where greatest dimension is less than 300.0 [12.00]	±.30 [±.012]	±.20 [±.008]	±.10 [±.004]
Where greatest dimension is greater than 300.0 [12.00]	±.40 [±.016]	±.30 [±.012]	±.20 [±.008]

Table B-10   Feature Location Tolerances for Solder Mask Aperture

Related Board Size	Class A	Class B	Class C
Where greatest dimension is less than 300.0 [12.00]	±.40 [±.016]	±.25 [±.010]	±.15 [±.006]
Where greatest dimension is greater than 300.0 [12.00]	±.45 [±.018]	±.30 [±.012]	±.20 [±.008]

Table B-11   Conductor Width Tolerances

Feature	Class A	Class B	Class C
Without plating	+.10 −.15 [+.004] [−.006]	+.05 −.10 [+.002] [−.004]	+.03 −.05 [+.001] [−.002]
With plating	+.20 −.15 [+.008] [−.006]	+.10 −.10 [+.004] [−.004]	+.08 −.08 [+.003] [−.003]

8. Type 2, the Basic Dimension System, is discussed beginning with paragraph 3.3. The exact dimension for a feature size or location is established, along with permissible variations. The same types of dimensions covered under type 1, Nominal Dimensioning, are also discussed in datail (see Tables B-12 to B-16). Only one example will be provided. (See Figure B-1 and Table B-17 from IPC-D-300G for an example of how tolerance may be expressed.)

Table B-12   Tolerance for Basic Dimensions of Cutouts, Notches, and Keying Slots, as Machined

Tolerances to Be Applied To:	Class A	Class B	Class C
Feature (slot or notch).	.15 [.006]	.10 [.004]	.05 [.002]
Location where greatest basic location dimension is less than 300.0 [12.00]	.20 [.008]	.15 [.006]	.10 [.004]
Location where greatest basic location dimension is greater than 300.0 [12.00]	.25 [.010]	.20 [.008]	.15 [.006]

NOTE: Radii should be provided in all slots or notches.

Table B-13   Board Thickness Tolerances

Basic Code	Class A	Class B	Class C
T1	.40 [.016]	.20 [.008]	.10 [.004]
T2	.60 [.024]	.40 [.016]	.20 [.008]
T3	.80 [.030]	.50 [.020]	.30 [.012]
T4	30%	20%	10%

Table B-14   Hole Location Tolerances

Related Board Size	Class A	Class B	Class C
Where greatest basic dimennsion is less than 300.0 [12.00]	.40 [.016]	.30 [.012]	.15 [.006]
Where greatest basic dimension is greater than 300.0 [12.00]	.55 [.022]	.40 [.016]	.30 [.012]

Table B-15   Feature Location Tolerances (Lands, Conductor Pattern, Etc.)

Related Board Size	Class A	Class B	Class C
Where greatest dimension is less than 300.0 [12.00]	.85 [.034]	.55 [.022]	.30 [.012]
Where greatest dimension is greater than 300.0 [12.00]	1.15 [.046]	.85 [.034]	.55 [.022]

NOTE: Conductor pattern registration may be expressed in terms of minimum annular ring violation, which establishes manufacturing registration allowances.

Table B-16   Feature Location Tolerances for Solder Mask Apertures

Related Board Size	Class A	Class B	Class C
Where greatest dimension is less than 300.0 [12.00]	1.15 [.046]	.70 [.028]	.40 [.016]
Where greatest dimension is greater than 300.0 [12.00]	1.30 [.052]	.85 [.034]	.55 [.022]

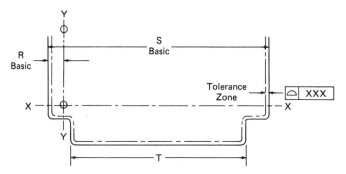

Fig. B-1   Tolerancing at contact finger tabs.

Table B-17   Board Edge Tolerance (Profile Tolerancing)

Class A	Class B	Class C
.40	.25	.15
[.016]	[.010]	[.005]

Table B-18   Summary of the Type 1 Dimensioning System

Applicable Characteristics	Paragraph Number	Figure Number	Table Number	Range of Dimensional Requirements
1. Board edges	3.2.1	6	4	.15–.40 [0.006–0.016]
2. Board connector tang	3.2.1.1	6	—	—
3. Board thickness	3.2.2	—	6	.05–.40 [.002–.016]
4. Board edge chamfering	3.2.2.1	8	—	—
5. Bow and twist (flatness)	3.1.4	—	3a, 3b, 3c	.5%–2.5%
6. Standard manufacturing allowances	3.1.3.2	—	2	.30–.85 [.012–.034]
7. Annular ring	3.1.3.1	—	1	.03–.40 [.001–.016]
8. Cutouts, notches, and keys	3.2.1.2	7	5	.05–.25 [.002–010]
9. Unsupported holes	3.2.3	—	7	.03–.15 [.001–.006]
10. Plated-through holes	3.2.4	—	8	.05–.20 [.002–.008]
11. Hole locating tolerances	3.2.5	5	9	.05–.20 [.002–.008]
12. Conductive pattern tolerances	3.2.6	—	10	.10–.40 [.009–.016]
13. Conductor width tolerances	3.2.8	—	12	.03–.20 [.001–.008]
14. Solder mask apertures	3.2.7	—	11	.15–.45 [.006–.018]

SECTION 4.0 SUMMARY

Tables B-18 and B-19 provide summaries of each dimensioning system. The paragraph and applicable table number are referenced. Since these table do not list all classes, most of the tables from IPC-D-300G are also reproduced.

SECTION 50.0 APPENDIX

1. Geometric characters and symbols (see Figures B-2 and B-3).
2. Hole-to-land relationship (50.2).
3. Detailed discussion of the feature size determination equation (50.2.1)
4. Conversion table of coordinate tolerance to positional (circular) tolerance.

Table B-19   Summary of the Type 2 Dimensioning System

Applicable Characteristics	Paragraph Number	Figure Number	Table Number	Range of Dimensional Requirements
1. Board edges	3.3.1	9	13	.15–40 [.006–.016]
2. Board connector tang	3.3.1.1	9	—	—
3. Board thickness	3.3.2	—	15	.10–.80 [.004–.030]
4. Board edge chamfering	3.2.2.1	8	—	—
5. Bow and twist (flatness)	3.1.4	—	3a, 3b, 3c	.5%–2.5%
6. Standard manufacturing allowances	3.1.3.2	—	2	.30–.85 [.012–.034]
7. Annular ring	3.1.3.1	—	1	.03–.40 [.001–.016]
8. Cutouts, notches, and keys	3.3.1.2	7	14	.05–.25 [.002–.010]
9. Unsupported holes	3.3.3	—	7	.03–.15 [.001–.006]
10. Plated-through holes	3.3.4	—	8	.05–.20 [.002–.008]
11. Hole locating tolerances	3.3.5	—	16	.15–.55 [.006–.022]
12. Conductive pattern tolerances	3.3.6	—	17	.30–1.15 [.012–.046]
13. Conductor width tolerances	3.3.8	—	12	.03–.20 [.001–.008]
14. Solder mask apertures	3.3.7	—	18	.40–1.30 [.016–.052]

"For editoral reasons certain tables and figures listed in *(name of document goes here* has been re-numbered. The reader is advised to obtain copies of those military and IPC documents."

	TYPE OF TOLERANCE	CHARACTERISTIC	SYMBOL
FOR INDIVIDUAL FEATURES	FORM	Straightness	—
		Flatness	▱
		Circularity (roundness)	○
		Cylindricity	○
FOR INDIVIDUAL OR RELATED FEATURES	PROFILE	Profile of a line	⌒
		Profile of a surface	⌓
FOR RELATED FEATURES	ORIENTATION	Angularity	∠
		Perpendicularity	⊥
		Parallelism	//
	LOCATION	Position	⊕
		Concentricity	◎
	RUNOUT	Circular runout	↗*
		Total runout	⫽↗*
*Arrowhead(s) may be filled in.			

Fig. B-2   Geometric characteristic symbols.

Term	Abbreviation	Symbol
At maximum material condition	MMC	Ⓜ
Regardless of feature size	RFS	Ⓢ
At least material condition	LMC	Ⓛ
Projected tolerance zone	TOL ZONE PROJ	Ⓟ
Diameter	DIA	⌀
Spherical diameter	SPHER DIA	S⌀
Radius	R	R
Spherical radius	SPHER R	SR
Reference	REF	( )
Arc length	A/C	⌒
All around	A/A	⊖⌀

Fig. B-3   Other symbols.

5. Microsectioning procedure
6. Thickness, plating in holes, micro-ohms method
7. Method for measuring bow and twist

Appendix C

MIL-P-55110D: General Specification for Printed Wiring Boards

This is an important document for a number of reasons which will become apparent. All printed circuit planning, manufacturing, and quality assurance personnel should have an understanding of and familiarity of its requirements. MIL-P-55110D and IPC-A-600C are perhaps the best presentation on end product acceptability and inspection guidelines available. IPC-A-600C is covered in Appendix D. No quality or planning persons should consider themselves competent and knowledgeable without having learned the information contained in both of these documents.

MIL-P-55110D also has another purpose. It stipulates that printed circuit boards being procured under government contract shall be procured only from manufacturers who have been qualified by the Defense Electronics Supply Center (DESC) as having met all documentation, manufacturing, and testing requirements listed in Table C-1. This document defines the exact qualification procedure to be followed. See paragraphs 4.5 and 6.6.

SECTION 1. SCOPE

This section states that this specification establishes the qualification and performance requirements of rigid single-sided, double-sided, and multilayer printed-wiring boards with plated-through holes. It also establishes three classifications and defines them as follows:

Type 1: Single-sided boards
Type 2: Double-sided boards
Type 3: Multilayer boards

Table C-1 Qualification Inspection

Inspection	Requirement Paragraph	Method Paragraph	Qualification Test Specimen Number (see 4.5.1)	Test Coupon by Board Type[1] 1	2	3	Whole Specimen
Material	3.4, 3.4.1– 3.4.9	—	—	Manufacturer certification			—
Visual							
Edges of laminate	3.5.1	4.8.2.1	1, 2, 3, 4	—	—	—	X
Surface imperfections	3.5.2	4.8.2.2	1, 2, 3, 4	—	—	—	X
Subsurface imperfections	3.5.3	4.8.2.3	1, 2, 3, 4	—	—	—	X
Marking	3.5.4	4.8.2.4	1, 2, 3, 4	—	—	—	X
Traceability	3.5.4.1	4.8.2.4	1, 2, 3, 4	—	—	—	X
Workmanship	3.5.5	4.8.2.5	1, 2, 3, 4	—	—	—	X
Solderability	3.5.6	4.8.2.6	—	—	—	—	—
Surface	3.5.6.1	4.8.2.6.1	1	A-1 B-1 C-1	A-1 B-1 C-1	A-1 B-1 C-1	—
Hole	3.5.6.2	4.8.2.6.2	1	—	E-1	E-1	—
Thermal stress	3.5.7	4.8.2.7	1	B-3	—	—	—
Dimensional	3.6, 3.6.1	4.8.3					
Hole pattern	3.6.2	4.8.3.1		—	—	—	X
Bow and twist	3.6.3	4.8.3.2	1, 2, 4	—	—	—	X
Conductor spacing	3.6.4	4.8.3.3	1, 2, 4	E-1 to E-5	E-1 to E-5	E-1 to E-5	—
Conductor pattern	3.6.5	4.8.3.4	1, 2, 4	E-1 to E-5	E-1 to E-5	E-1 to E-5	X
Layer-to-layer registration	3.6.6	4.8.3.5	1, 2, 3, 4	—	—	—	X

Table C-1 (Continued)

Inspection	Requirement Paragraph	Method Paragraph	Qualification Test Specimen Number (see 4.5.1)	Test Coupon by Board Type[1]			Whole Specimen
				1	2	3	
Dimensional (Continued)							
Annular ring (external)	3.6.7	4.8.3.6	1, 4	A-3	—	—	X
Unsupported hole	3.6.7.1	4.8.3.6	1, 4	A-3	—	—	X
Plated-through hole	3.6.7.2	4.8.3.6	1, 4	—	A	A	X
Plating and coating thickness	3.6.9	4.8.3.8	3	C-1, C-4, C-5	C-1, C-4, C-5	C-1, C-4, C-5	—
Physical requirements	3.7	4.8.4	—	—	—	—	—
Plating adhesion	3.7.2	4.8.4.2	1, 2, 3, 4	C	C	C	X
Conductor edge outgrowth	3.7.3	4.8.4.3	—	—	—	—	X
Bond strength	3.7.4	4.8.4.4	1	A-2	—	—	—
Construction integrity (microsection)	3.8	4.8.5	1, 4	—	—	—	—
Plated-through hole	3.8.1	4.8.5.1	1, 4	—	A-1 or A-5	A-1 or A-5	—
Plated copper thickness	3.8.2	4.8.5.2	1, 4	—	A-1 or A-5	A-1 or A-5	—
Layer to layer[2]	3.6.6	4.8.3.5	1, 4	—	A-1 or A-5	A-1 or A-5	—
Plating voids	3.8.3	4.8.5.3	1, 4	—	A-1 or A-5	A-1 or A-5	—
Conductor thickness	3.8.4	4.8.5.4	1, 4	—	A-1 or A-5	A-1 or A-5	—
Resin smear and etchback	3.8.5	4.8.5.5	—	—	—	—	—
Hole cleaning (smear removal)	3.8.5.1	4.8.5.5	1, 4	—	—	A-1 or A-5	—
Negative etchback	3.8.5.2	4.8.5.5	1, 4	—	—	A-1 or A-5	—
Etchback	3.8.5.3	4.8.5.5	1, 4	—	—	A-1 or A-5	—

Undercutting	3.8.6	4.8.5.6	1, 4	—	A-1 or A-5	A-1 or A-5	—
Annular ring (internal)	3.8.7	4.8.5.7	1, 4	—	—	A-1 or A-5	—
Dielectric layer thickness	3.8.8	4.8.5.8	1, 4	—	A-1 or A-5	A-1 or A-5	—
Laminate voids	3.8.9	4.8.5.9	1, 4	—	A-1 or A-5	A-1 or A-5	—
Resin recession	3.8.10	4.8.5.10	1, 4	—	A-1 or A5	A-1 or A-5	—
Lifted lands	3.8.11	4.8.5.11	1, 4	—	A-1 or A-5	A-1 or A-5	—
Plated-through holes	3.9	4.8.6	—	—	—	—	—
Thermal stress	3.9.1	4.8.6.1	1	—	B-2	B-2	—
Layer to layer[2]	3.6.6	4.8.3.5	1	—	—	B-2	—
Rework simulation	3.9.2	4.8.6.2	1, 2	—	A-2	A-2	—
Thermal shock	3.9.3	4.8.6.3	1, 2	—	A-2	A-2	—
Lifted lands	3.9.4	4.8.6.4	1	—	D-3	D-3	—
Electrical and environmental requirements	3.10	4.8.7	—	—	—	—	—
Moisture and insulation resistance	3.10.1	4.8.7.1	1	E-1	E-1	E-1	—
Dielectric withstanding voltage	3.10.2	4.8.7.2	1	E-1	E-1	E-1	—
Circuitry	3.10.3	4.8.7.3	—	—	—	—	—
Circuitry continuity	3.10.3.1	4.8.7.3.1	2	D-3	D-3	D-3	—
Circuit shorts[3]	3.10.3.2	4.8.7.3.2	2	E-1	E-1	E-1	—
Cleanliness[4]	3.10.4	4.8.7.4	1, 2, 4	All	All	All	X
Ionic	3.10.4.1	4.8.7.4.1	1, 2, 4	All	All	All	X
Repair	3.10.5	4.8.7.5	1, 2, 3, 4	All	All	All	X

[1] See MIL-STD-275, paragraph 1.2 and Figures 11, 12, and 13 herein.
[2] The two layer-to-layer microsections shall be at 90 degree angles.
[3] 200 Volts.
[4] Cleanliness shall be performed prior to any other inspection.

SECTION 2. APPLICABLE DOCUMENTS

The purpose of these other documents is as discussed in MIL-STD-275E under "Referenced Documents." The planning and quality departments should have a copy of these documents on hand for reference as needed. Without these documents, there is no way to verify a requirement which may arise on any job.

SECTION 3. REQUIREMENTS

This section details all manufacturing and other requirements for rigid printed circuit boards. They are discussed below.

1. Quality conformance test coupons shall be included on all panels. Defects noted in the coupons shall be considered indicative of defects in the printed circuit boards themselves, and corrective action must be taken.

2. Metal-clad laminates used are to be in accordance with MIL-P-13949 (plastic sheet, metal-clad, for use in printed wiring boards):
   a. No base material (dielectric) shall be used which is thinner than .002 inch, not including copper foil, for single-sided cladding.
   b. Minimum base material width with double-sided cladding shall be .0035 inch.

   Other materials such as fluxes, copper foil, solder, solder mask, marking inks, and adhesives are specified in paragraph 3.4.

3. Visual requirements are discussed in paragraph 3.5. A very brief and only partial summary of them is provided below. All of the requirements in paragraph 3.5 should be studied and known well by planners and quality assurance personnel.
   a. Haloing along board edges shall be limited to .100 inch or 50% of the edge spacing called out on the drawing.
   b. Surface imperfections (such as scratches, dents, and haloing) are allowed, provided that laminate fiber is not damaged or exposed and the imperfection does not bridge conductors or reduce dielectric spacing requirements.
   c. Subsurface imperfections (blistering, haloing, measles) are allowed, provided that the imperfection is nonconductive, does not bridge more than 25% of the distance between conductors, does not reduce conductor spacing below requirements, and does not propagate as a result of testing.
   d. Foreign particles included in laminate are allowed, provided they are at least .010 inch from conductors, do not reduce spacing more than 50%,

are smaller than .032 inch, and are limited to a maximum of two such defects per side of the printed circuit board. Gelation particles (pieces of epoxy resin from prepreg) are permitted, regardless of their location.

e. Measling is allowed within the restricted guidelines of paragraph 3.5.3.3.

f. All boards and quality conformance coupon strips should be marked per the drawing and contain the following:

(1) Part number and revision level.

(2) Date of manufacture.

(3) Manufacturer's Federal Supply Code for Manufacturers (FSCM).

(4) Lot number.

(5) Serial number, which identifies each coupon and board made on a given panel.

NOTE: Markings A, B, C, and D can be etched/plated on by inclusion with the outer layer artwork during imaging, or they can be applied by an appropriate ink.

NOTE: The serial number must be stamped on with an appropriate ink or scribed onto a metal square provided for that purpose. It is required that all boards and coupons be traceable to the panels from which they were manufactured.

g. The boards shall exhibit good workmanship when inspected in accordance with other requirements and methods called out in this specification.

h. The boards must meet minimum solderability requirements.

i. Dimensional requirements and hole pattern accuracy shall meet the drawing specifications.

j. Maximum bow and twist shall be 1.5% when tested according to the method described in paragraph 4.8.3.2.

k. Conductor width and spacing shall be per the drawing. If nothing is specified on the drawing, minimum external spacing shall be .005 inch and minimum internal spacing .004 inch. Internal or external conductor width should not be less than .004 inch.

l. Conductor pattern defects shall not reduce conductors more than 20% below the minimum width requirement on the drawing. These isolated defects must be confined to a .500-inch section or less of a conductor's length.

NOTE: Thus, isolated defects which violate the minimum conductor width requirements may be permissible.

m. Layer-to-layer registration cannot deviate more than .014 inch for a multilayer board. Registration must be confirmed by two mutually per-

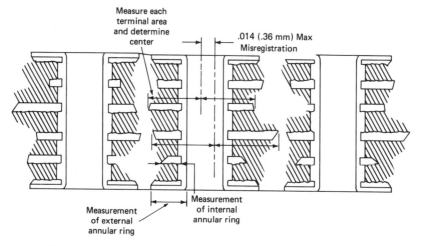

Fig. C-1 Layer-to-layer registration and annular ring measurement.

pendicular microsections or by special coupons which allow visual assessment (see Figure C-1).

n. Minimum annular ring requirements may be violated by up to 20% due to isolated defects on external layers. Other requirements for the minimum annular ring are as follows:

(1) External plated-through hole: .005 inch where the conductor meets the hole; .002 inch elsewhere; unsupported hole: .015 inch.

(2) Internal: .002 inch (see Figure C-2).

o. Solder mask thickness at the crest of a conductor shall be .001 inch minimum. It is permissible to have a certain amount of solder mask peel off when tape tested, unless that mask is on bare copper or bare laminate (see Table C-2). If solder mask is required, the boards are to be tested for ionic contamination prior to applying the solder mask.

p. Plating thicknesses shall be per Table C-3. Note that tin-lead is to be measured before reflow not afterward.

q. Construction integrity shall be determined by microsection prior to thermal stress testing. The requirements are listed in paragraph 4.8.5 (see Figures C-3 to Figure C-9).

The plated-through hole is examined by microsectioning of three holes in a coupon section. Good workmanship should be evident and the following requirements apply:

(a) No cracks in conductive foil, plating, or coating.

(b) No separation at conductor interfaces.

(c) Nail heading shall not exceed 1.5 times the foil thickness.

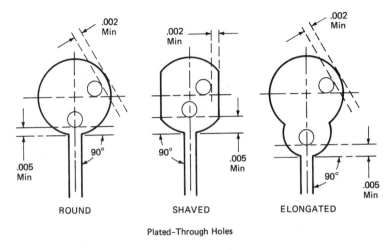

Plated-Through Holes

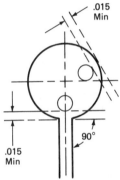

Unsupported Holes

Fig. C-2   Land areas (external minimum annular ring).

Table C-2   Solder Mask Adhesion to Printed Wiring Boards

Material	Maximum Percentage of Lifting: Melting and Nonmelting Metals Nonscribed Test
Bare copper	0
Gold or nickel	5
Tin-lead plating	10
Reflowed tin-lead	10
Base laminate	0

Table C-3 Plating and Coating Thickness[1]

Plating Material	Surface and Through-Hole Thickness
Gold	.000050 inch minimum
Nickel	.0002 inch minimum
Tin-lead	.0003 inch minimum at the surface as plated
Solder coating	.0003 inch minimum at the crest on the surface as coated

[1]A coupon prior to reflow may be required (see paragraph 4.6 and Table C-7).

(d) Nodules, plating folds, or plated glass fiber protrusion are acceptable, provided that hole diameter and copper thickness requirements are not violated.

r. Surface and through-hole copper thickness shall be .001 inch minimum, except that isolated areas down to .0008 inch are permissible. Isolated areas less than .0008 inch shall be treated as a void.

s. Plating voids in the hole wall require 100% visual inspection. Allowable voids are limited to the following (three voids maximum in a hole):

(1) The combined length of voids shall not exceed 5% of the total hole wall length.

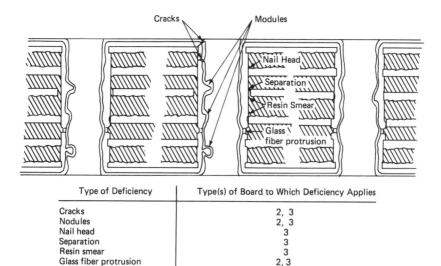

Type of Deficiency	Type(s) of Board to Which Deficiency Applies
Cracks	2, 3
Nodules	2, 3
Nail head	3
Separation	3
Resin smear	3
Glass fiber protrusion	2, 3

Fig. C-3 Deficiencies in plated-through hole workmanship.

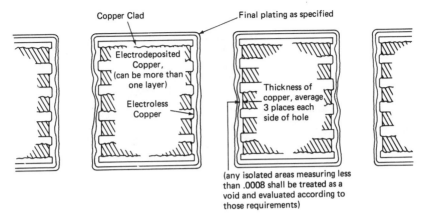

Fig. C-4 Plating thickness.

(2) The combined area shall not exceed 10% of the hole wall area.

(3) No circumferential voids (ring voids or lip voids) are permitted.

t. Etchback and smear removal requirements

(a) Type 3 boards shall be free of resin smear at inner layer connections.

(b) When etchback is not specified, lateral removal of hole wall material (etchback) shall not exceed .001 inch.

(c) When etchback is specified on the drawing, lateral removal of hole wall material shall be .0002 inch minimum to .003 inch maximum. Material removal may take place on one or both sides of the inner layer foil. Wicking or copper back an additional .003 inch maximum is allowable, provided that conductor spacing requirements are not violated.

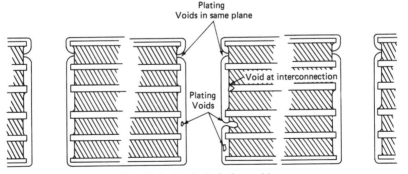

Fig. C-5 Typical plating voids.

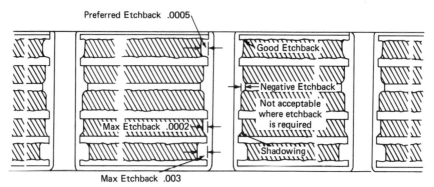

Preferred Etchback .0005

Good Etchback

Negative Etchback
Not acceptable
where etchback
is required

Max Etchback .0002

Shadowing

Max Etchback .003

NOTES:
1. Dimensions are in inches.

Fig. C-6   Forms of etchback.

u.   Negative etchback is the lateral removal of inner layer copper foil at the plated-through hole. It should not exceed .0005 inch.

v.   Undercutting shall not exceed the total thickness of foil and plating or 10% of the conductor width, whichever is less.

w.   Dielectric layer thickness shall be per the drawing. Type 3 boards shall have .0035-inch minimum spacing between any two conductive layers (see Figure C-7). There must be a minimum of two sheets of prepreg or laminate between any two conductive layers.

x.   Laminate voids .003 inch or less in the longest direction shall be permitted.

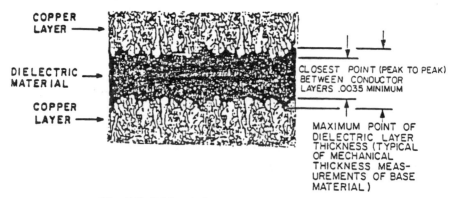

COPPER
LAYER

DIELECTRIC
MATERIAL

COPPER
LAYER

CLOSEST POINT (PEAK TO PEAK)
BETWEEN CONDUCTOR
LAYERS .0035 MINIMUM

MAXIMUM POINT OF
DIELECTRIC LAYER
THICKNESS (TYPICAL
OF MECHANICAL
THICKNESS MEAS-
UREMENTS OF BASE
MATERIAL)

Fig. C-7   Dielectric layer thickness measurement.

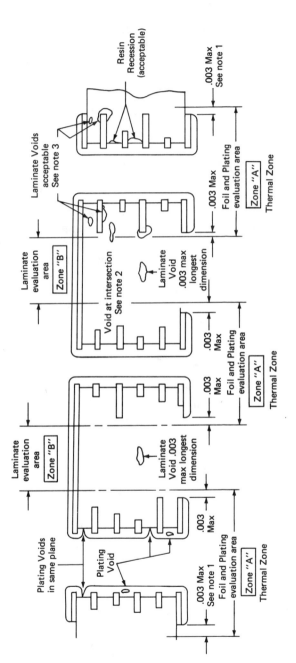

NOTES:

1. Typically beyond land edge most radially extended.
2. Void at intersection of Zone A and Zone B. Laminate voids greater than .003 (0.08 mm) in length which extend into the laminate evaluation area are rejectable.
3. Laminate voids are not evaluated in Zone A, laminate voids greater than .003 (0.08 mm) that extend into Zone B are rejectable.
4. Dimensions are in inches.

Fig. C-8   Typical microsection of plated-through holes after thermal stress and rework simulation.

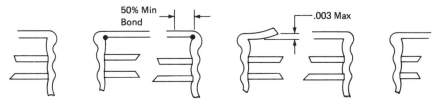

NOTES:
1.   Dimensions are in inches.
2.   Metric equivalents are given for general information only.

Fig. C-9   Lifted lands.

y.   Resin recession from the hole wall shall be permitted as long as it is less than 40% of the hole wall length and does not recede more than .003 inch from the hole wall copper in the unthermal stress sample. Resin recession exceeding these requirements is permitted in the thermal stressed sample.

z.   There shall be no lifted land (metal areas) in the microsection sample prior to thermal stress. After thermal stress, lifting is allowed, provided at least 50% of the land is bonded and the lifted portion is less than .003 inch off the surface.

aa.   Thermal stressed microsections shall be examined in two zones, A and B. (see Figure C-8).

bb.   Electrical continuity testing shall be performed on all production boards.

cc.   Repair is not permitted on boards being built to MIL-P-55110D requirements. When boards are inspected, there shall be no evidence of repair.

NOTE: Touchup is permitted. Touchup is defined as repeating a manufacturing operation manually to increase the yield on acceptable boards.

SECTION 4. QUALITY ASSURANCE PROVISIONS

This section spells out all inspection and testing requirements needed to operate a MIL-P-55110D program. There are four types of inspection: (1) materials inspection, (2) qualification inspection, (3) in-process inspection, (4) quality conformance inspection (groups A and B).

1.   Materials inspection. This consists of certifications from the manufacturers together with verifying data. The verifying data (laboratory/inspection reports) must be available upon request. Table C-4 lists those items which must be covered by this requirement and the specification which they must meet.

Table C-4   Materials Inspection

Material	Requirement Paragraph	Applicable Specification
Metal-clad laminate	3.4.1	MIL-P-13949
Bonding material	3.4.2	MIL-P-13949
Solder coating	3.4.4	QQ-S-571
Soldering flux	3.4.5.3	MIL-F-14256
Permanent solder mask	3.4.6	IPC-SM-840
Copper foil	3.4.3	IPC-CF-150

2. Qualification inspection. This is performed at a laboratory acceptable to the government and is part of the procedure for becoming a DESC qualified manufacturer of printed circuit boards. Table C-1 lists all requirements. The entire qualification program is discussed beginning with paragraph 4.5.

3. In-process inspection. The requirements are listed in Table C-5. Most of them are met simply by performing group A inspections (see Table C-6). However, tin-lead thickness must be measured and recorded as the job is being plated. Also, a minimum of five boards per shift must have ionic contamination testing performed. This testing shall be considered an ongoing program required of all DESC qualified producers (see Table C-5).

4. Quality conformance inspection. This is the group A and group B listed in Tables C-6 and C-7. Group A must be performed on each lot of boards being run. Group B must be performed, as a minimum, on the most complicated pattern of printed circuits produced during a month. For group B, two coupon strips and the associated boards are submitted to a government-approved laboratory for testing.

SECTION 5. PACKAGING

Three levels of preservation are discusssed: A, B, and C.

Table C-5   In-Process Inspection

Test	Requirement Paragraph	Method Paragraph
Cleanliness	3.10.4, 3.10.4.1	4.8.7.4, 4.8.7.4.1
Plating deposit[1]	3.4.3 and 3.6.9	4.8.1, 4.8.3.8
Solder mask	3.4.6	4.8.3.7
Conductor pattern	3.6.5	4.8.3.4
Plating adhesion	3.7.2	4.8.4.2

[1]A nonreflowed coupon may be required by contract.

Table C-6   Group A Inspection

Inspection	Requirement Paragraph	Method Paragraph	Production Board	Test Coupon by Board Type[1]			AQL (Percent Defective)	
				1	2	3	Major	Minor
Material	3.4, 3.4.1 to 3.4.7	—	—	Manufacturer certification			—	—
Visual	3.5	4.8.2	[2]					
Edges of printed wiring board	3.5.1	4.8.2.1	X	—	—	—	1.0	4.0
Surface imperfections	3.5.2	4.8.2.2	X	—	—	—	1.0	4.0
Subsurface imperfections	3.5.3	4.8.2.3	X	—	—	—	1.0	4.0
Marking	3.5.4	4.8.2.4	X	—	—	—	1.0	4.0
Traceability	3.5.4.1	4.8.2.4	X	—	—	—	1.0	4.0
Workmanship	3.5.5	4.8.2.5	X	—	—	—	1.0	4.0
Solderability	3.5.6	4.8.2.6	—[3]	—	—	—	—	—
Surface	3.5.6.1	4.8.2.6.1	—	C	—	—	1.0	4.0
Hole	3.5.6.2	4.8.2.6.2	—	—	A	A	1.1	4.0
Thermal stress	3.5.7	4.8.2.7	—	B	A	A	1.0[6]	4.0
Dimensional	3.6, 3.6.1	4.8.3	—	—	—	—	1.0	4.0
Hole pattern	3.6.2	4.8.3.1	X	—	—	—	1.0	4.0
Bow and twist	3.6.3	4.8.3.2	X	—	—	—	1.0	4.0
Conductor spacing	3.6.4	4.8.3.3	X[7]	—	—	—	1.0	4.0
Conductor pattern	3.6.5	4.8.3.4	—	—	—	F	1.0	4.0
Layer-to-layer registration	3.6.6	4.8.3.5	X	—	[8]	—	1.0	4.0
Annular ring (external)	3.6.7	4.8.3.6	X				1.0	4.0
Unsupported hole	3.6.7.1	4.8.3.6	X	—	—	—	1.0	4.0
Plated-through hole	3.6.7.2	4.8.3.6	X				1.0	4.0

Requirement	Req. para.	Test method						
Solder mask thickness	3.6.8	4.8.3.7	X³	E³	E³	E³	1.0	4.0
Plating and coating thickness	3.6.9	4.8.3.8	X³	C³	C³	C³	1.0	4.0
Physical requirements	3.7	4.8.4	—	—	—	—	—	—
Solder mask cure and adhesion	3.7.1	4.8.4.1	X³	J³	J³	J³	1.0	4.0
Plating adhesion	3.7.2	4.8.4.2	X³	C³	C³	C³	1.0	4.0
Conductor edge outgrowth[9]	3.7.3	4.8.4.3	X	—	—	—	1.0	4.0
Construction integrity (microsection)[4]	3.8	4.8.5	—	—	—	—	—	—
Plated-through hole	3.8.1	4.8.5.1	—	—	B	B	4	4
Plated copper thickness	3.8.2	4.8.5.2	—	B	B	B	4	4
Plating voids	3.8.3	4.8.5.3	—	B	B	B	4	4
Conductor thickness	3.8.4	4.8.5.4	—	B	B	B	4	4
Resin smear and etchback	3.8.5	4.8.5.5	—	—	—	—	—	—
Hole cleaning (smear removal)	3.8.5.1	4.8.5.5	—	—	—	B	4	4
Negative etchback	3.8.5.2	4.8.5.5	—	—	—	B	4	4
Etchback	3.8.5.3	4.8.5.5	—	—	—	B	4	4
Undercutting	3.8.6	4.8.5.6	—	B	B	B	4	4
Annular ring (internal)	3.8.7	4.8.5.7	—	—	B	B	4	4
Dielectric layer thickness	3.8.8	4.8.5.8	—	—	B	B	4	4
Laminate voids	3.8.9	4.8.5.9	—	—	B	B	4	4
Resin recession	3.8.10	4.8.5.10	—	B	B	B	4	4
Lifted lands	3.8.11	4.8.5.11	—	—	B	B	4	4
Plated-through holes	3.9	4.8.6	—	—	—	—	5	—
Thermal stress	3.9.1	4.8.6.1, 4.8.1	—	B⁴	B⁴	B⁴	—	5
Electrical and environmental requirements	3.10	4.8.7	—	—	—	—	—	—

Table C-6   (Continued)

Inspection	Requirement Paragraph	Method Paragraph	Production Board	Test Coupon by Board Type[1]			AQL (Percent Defective)	
				1	2	3	Major	Minor
Circuitry	3.10.3	4.8.7.3	—	—	—	—	—	—
Circuitry continuity	3.10.3.2	4.8.7.3.1	X	—	—	—		100% inspection[10]
Circuit shorts	3.10.3.3	4.8.7.3.2	X	—	—	All		100% Inspection[10]
Repair	3.10.5	4.8.7.5	X	All	All	All		100% Inspection

[1] See MIL-STD-275 and paragraph 1.2.

[2] Visual examination (4.8.1) of production board surface for all three board types (1, 2, and 3).

[3] Test coupon or production board; manufacturer's option coupon shall be processed with production board.

[4] One coupon per panel shall be microsectioned for type 3 boards; the number of coupons to be microsectioned for types 1 and 2 boards shall be based on a statistical sample in accordance with MIL-STD-105 General Inspection level II of the number of panels produced and shall meet an AQL of 2.5% defective.

[5] For type 3 boards, microsection 1 coupon per panel 100% of the time in any one direction, and microsection perpendicular to that direction on a sampling of the microsectioned coupons based on MIL-STD-105 General Inspection level II with an AQL of 2.5% defective. Type 2 boards shall be microsectioned in only one direction.

[6] See 4.7.1.2 of MIL-P-55110.

[7] Inspected prior to lamination.

[8] Production board shall be used for type 2.

[9] May be inspected by examination of microsectioned coupon associated with production board.

[10] If the printed assembly drawing required the circuitry tests to be run with 100% inspection on the printed wiring assembly, a sampling plan (4.7.1.2.1) based on an AQL of 2.5% defective shall be used on the bare unassembled printed wiring board.

Table C-7   Group B Inspection

Inspection	Requirement Paragraph	Method Paragraph	Test Coupon by Type[1]		
			1	2	3
Bond strength	3.7.4	4.8.4.4	B	—	—
Rework simulation	3.9.2	4.8.6.2	—	B	B
Moisture and insulation resistance	3.10.1	4.8.7.1	E	E	E
Dielectric withstanding voltage	3.10.2	4.8.7.2	E	E	E

[1]See MIL-STD-275 and 1.2 herein.

SECTION 6. NOTES

This section has isolated pieces of information which supplement the other sections. Included are definitions and explanations.

Appendix D

IPC-A-600C: Guidelines for Acceptability of Printed Boards

Technically this document is part of inspection literature for quality assurance. It was compiled to help "standardize individual interpretations to specifications on printed boards." Because of the need to place standards and specifications within the visual context of the end product, it is desirable that design, quality, planning, and manufacturing personnel know what certain conditions look like and how those conditions are viewed by the electronics/printed circuit industry (i.e., IPC members). Without a knowledge of this document and the other standards and specifications in this chapter, a design, planning, quality assurance, or manufacturing person would have no basis for evaluating a company specification, a set of artwork, or a set of blueprints for a given printed circuit board.

This document establishes three conditions: preferred, acceptable, and rejectable. The meaning of "preferred" is obvious. "Acceptable" means that the product is reliable and functionable, although poor workmanship is evident. "Rejectable" means that the product has a very poor aesthetic appearance and may not meet reliability requirements.

IPC-A-600C is divided into 15 sections. Each section should be read and understood by people involved in any aspect of printed circuit manufacturing. This document is perhaps the most well-known and often-cited one in printed circuit literature. It forms part of the internal specifications for many government and industrial organizations (see Table D-1).

SECTION 1. PLATED-THROUGH HOLES

This section contains drawings and photographs of microsectioned holes showing numerous conditions related to the acceptability of the plated-through hole. It is prefaced by a list of methods used for inspecting and measuring. This list is important because it establishes certain methods and techniques as standard.

SECTION 2. SURFACE PLATING

1. Adhesion is tape tested. The only acceptable condition is that no metal is removed by the tape. There is a comment that slivering is often associated

Table D-1   Table of Contents to IPC-A-600C

with plating adhesion. Slivering is the breaking off of metallic etch resist along conductor edges.

2. Nodules (burned-on excessive plating) are acceptable, provided they are not loose and easily broken off, and provided minimum conductor spacing requirements are not violated.

3. Pitting should be cause for rejection if pits are large, numerous, and omnipresent or if they result in exposure of the base metal. Except in these situations, acceptability should be based on the intended function of the end product. A distinction is made between pitting in contact fingers and pitting elsewhere in the circuitry. This section makes it clear that pitted plating is not an automatic cause for rejection; accept/reject decisions should be based instead of functionality and intended use.

SECTION 3. SOLDER COATING

1. Acceptability is based on the ability of the coating to provide a reliable electrical and mechanical joint. The solder also provides some protection to the copper surface. Visual examination is the only requirement. It is usually performed after the edge dip solderability test called out in paragraph 2.4.14 of IPC-TM-650.

2. Dewetting is allowed, provided it is not widespread.

3. The thickness of fused solder is measured at the crest of the conductor.

   NOTE: IPC-A-600C does not address the quality requirements of SMOBC circuitry. These requirements are discussed in another separate chapter of this book.

SECTION 4. LAMINATE BASE MATERIAL

This section contains a detailed discussion of laminate defects and conditions. There are some important areas which have been neglected, such as discussions of included particles, damaged laminate, stains or discolorations, and darkened laminate as a result of baking.

1. Measles and crazing. There is an extensive discussion of these defects. Measles are white dots caused by separation of the resin and the glass fibers; they generally result from thermal processing. Crazing is more severe and has the appearance of connected measles. These defects are generally considered to result from mechanical processing. Some rather technical guide-

lines are set up for determining the acceptability of these conditions (see Tables D-1, D-2, and D-3). A paragraph is also added which states that the IPC has no data showing that a board with even severe measles has ever failed due to this defect, even with extended use under harsh conditions (see Figure D-1).

TABLE D-2   Acceptability Guidelines for Measling and Crazing: Raw Material—Bare Laminate

Class I: Consumer	Class II: General— Industrial	Class III: High-Reliability Life Support
Measling and crazing shall not exceed 2% of the total usable area. (See Notes.)	Measling and crazing shall not exceed .5% of the total usable area. (See Notes.)	Measling and crazing shall not exceed .1% of the total usable area. (See Notes.)

NOTE 1: The area of measling or crazing is determined by combining the area of each measle or craze and dividing by the total area of the printed board. A separate determination is made for each side of the board.
NOTE 2: The referee test (destructive) to determine propagation of measling or crazing is to precondition the test specimen and solder float the specimen on a solder bath at a temperature of 288°C ± 6°C (550° ± 10°F) for a period of 10 seconds. Measling/crazing shall not propagate into a lesser class requirement or other defect category. (See IPC Test Method 2.6.8.)
Courtesy of the IPC.

Table D-3   Acceptability Guidelines for Measling and Crazing: Ball Printed Board (Completely Processed and Fused)

Class I: Consumer	Class II: General— Industrial	Class III: High-Reliability Life Support
Measling or crazing does not bridge electrically uncommon conductors.	Total area affected by measles or crazing shall not exceed 5% of the board area.	Total area affected by measles or crazing shall not exceed 1% of the board area.
Test for dielectric resistance: 10,000 megohms/ 500 VDC between conductive surfaces where measles occur (dry board). (See Notes.)	There shall be no more than 70% reduction in space between electrically uncommon conductors. (See Notes.)	There shall be no more than 25% reduction in space between electrically uncommon conductors. (See Notes.)

NOTE 1: Repairs to boards are permissible provided that the repaired board meets the above criteria.
NOTE 2: The area of measling or crazing is determined by combining the area of each measle or craze and dividing by the total area of the printed board. A separate determination is made for each side of the board.
NOTE 3: The referee test (destructive) to determine propagation of measling or crazing is to precondition the test specimen and solder float the specimen on a solder bath at a temperature of 260°C ± 6°C (500° ± 10°F) for a period of 5 seconds. Measling/crazing shall not propagate into a lesser class requirement or other defect category. (See IPC Test Method 2.6.8., except that the time and temperature shall be as shown above.)
Courtesy of the IPC.

Table D-4   Acceptability Guidelines for Measling and Crazing: Printed Board Assemblies

Class I: Consumer	Class II: General—Industrial	Class III: High-Reliability Life Support
The only criterion for measling and crazing is that the assembly is functional.	Total area affected by measles or crazing shall not exceed 10% of the board area.	Total area affected by measles or crazing shall not exceed 2% of the board area.
This may be determined through functional testing or dielectric resistance measurements. (See Notes.)	There shall be no more than 80% reduction in space between electrically un-common conductors. (See Notes.)	There shall be no more than 50% reduction in space between electrically un-common conductors. (See Notes.)

NOTE 1: Repairs to boards are permissible provided that the repaired board meets the above criteria.
NOTE 2: The area of measling or crazing is determined by combining the area of each measle or craze and dividing by the total area of the printed board. A separate determination is made for each side of the board.
Courtesy of the IPC.

The evidence to date is that even boards with severe MEASLES have functioned adequately over long periods of time and in harsh environments. In fact, the IPC has no data which shows that a board which is "MEASLED" (and not subjected to other more serious conditions) has failed.

The committee established 3 classes of measles, one as applied to consumer products, one for general industrial use, and one for high reliability: life support systems. These findings and deliberations are detailed on pages 24 and 25.

Fig. D-1   IPC statement on measles.

Definitions, drawings, and photographs are provided for other conditions, but, acceptability guidelines are left to other documents.

2. Blistering and delamination are the same, except that delamination is small, localized areas is called "blistering."

3. Weave texture and weave exposure. Weave texture that is noticeable at the laminate surface is not a problem. When the glass fibers are not completely covered by resin, weave exposure is the result and may be cause for concern.

4. Haloing consists of light areas in the laminate around machined areas of the board.

SECTION 5. ETCHING CHARACTERISTICS

This section defines the terms used to discuss and measure the cross-sectional characteristics of conductors. It is noted that the trace width will generally increase (called "outgrowth") when screened-on resist is used for pattern plating and that the conductor width will generally decrease (called "undercut") when photoresist imaging is used for pattern plating. Thus, conductors will have a different cross-sectional appearance depending on the choice of imaging method.

1. Etch factor. This is the ratio of total copper thickness to overhang. Etch factor = copper/overhang (see Figure D-2, top right-hand corner).
2. Minimum conductor width (MCW), overall conductor width (OCW), and designed conductor width (DCW) are shown in Figure D-3.
3. Numerous photographs of etching conditions are shown.

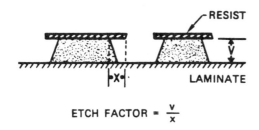

ETCH FACTOR = $\dfrac{v}{x}$

Fig. D-2 An etch factor of 1:1 is usually considered practical. Higher factors may be specified for some applications.

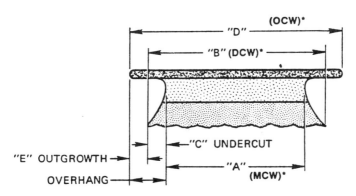

Fig. D-3 Cross section of a pattern plated conductor after dry film photoresist imaging.

SECTION 6. CONDUCTOR

This section defines conditions which may affect conductors, and presents drawings and photographs of these conditions. It also describes how these conditions are to be measured.

1. Various conditions are explained in Figures D-4 and D-5 from this documents. They include edge roughness, indentations, and isolated projections measured along a span of the conductor, usually a .500-inch section.

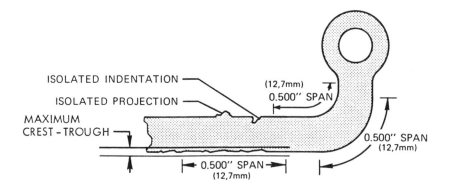

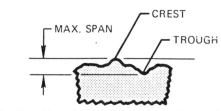

Fig. D-4   Irregular conditions on conductor edges.

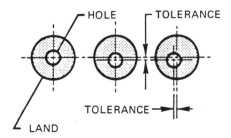

Fig. D-5   Misregistration.

2. Isolated projections are permissible as long as minimum conductor spacing is not violated.
3. Edge roughness is defined here but allowable limits are set by other documents, usually specified by contract or blueprint/procurements specifications.
4. Registration is determined by measuring the distance from the center of the hole to the center of the land area about that hole (see Figure D-5).
5. Conductor edge definition is defined as preferred, acceptable, or rejectable.

SECTION 7. FABRICATION

Drilling and machining of holes and edges are discussed, together with the appropriate measuring method. Preferred, acceptable, and rejectable conditions are depicted.

1. Land registration. Misregistration is acceptable provided minimum annular ring requirements are not violated.
2. Sawing/routing edge conditions. Only smooth edges are acceptable. No burrs are acceptable; they must be filed or sanded.
3. Solder mask registration. Solder mask on the pads is acceptable provided sufficient land is exposed to form an adequate solder fillet during soldering.

   NOTE: This is often interpreted to mean that a minimum annular ring requirement must be met for the land area exposed by the solder mask.

4. Chamfering (beveling along edges of contact fingers). Slight burrs are acceptable provided bridging due to burrs does not violate any specified conductor tolerance or spacing requirement. Lifted metal and loose glass fibers are rejectable, along with excessive burring.
5. Nonplated drill holes. Slight burrs and glass fibers are acceptable. Excessive degrees of these conditions indicate lack of process control and are rejectable.

SECTION 8. EYELETING

Acceptability guidelines, together with appropriate photographs, are displayed for funnel, flat, and rolled eyelets. There is a caution against using eyelets for through connections, as improper assembly may result in intermittent open circuits.

SECTION 9. LEGENDS

Legends may be either etched on or applied by screen printing or stamping. They must be legible and permanent and must not interfere with the function of the circuit in any way.

SECTION 10. FLUSH PRINTED BOARDS

The acceptability guidelines for these boards, which are identical to those for traditional printed circuit boards, are listed in this section by page number. Flush printed circuit conductors are rejectable if they are not flush, with the height above the base material or the surrounding insulating material exceeding allowable tolerances.

SECTION 11. FLEXIBLE PRINTED BOARDS

Guidelines are for these boards are similar to those for rigid boards. There is a set of photographs which show examples of delamination of the cover coat and the plated-through holes. Flexible and rigid-flex circuits are covered in Chapter 6 of this book.

SECTION 12. MULTILAYER PRINTED BOARDS

Most of the acceptability guidelines discussed thus far also apply to multilayer printed circuits. There are, however, additional concerns related to the plated-through holes and to inner layer integrity and registration. Coupons should be run with multilayer circuitry. This allows destructive testing (such as microsectioning) to be performed without having to sacrifice an otherwise acceptable circuit.

1. Integrity of the plating junction of the plated hole and inner layer must be assured by the use of chemical cleaning or etchback processing.
2. Minimum annular rings at plated holes and layer-to-layer registration are measured by X-ray photography or mutually perpendicular microsections.
3. Lamination integrity is verified by inspecting for air entrapment, layer-to-layer delamination, and localized blisters (delamination).
4. Electrical integrity is verified by checking for power-to-ground shorts; followed by continuity testing of the circuitry.
5. Photographs of preferred, acceptable, and rejectable conditions are shown.

SECTION 13. BOW AND TWIST

This section presents charts which show allowable variations for single, double, and multilayered circuitry according to the type of material and board thickness. The techniques to be used for determining bow and twist are shown. This section is reproduced here (see Figure D-6).

The board shall be placed unrestrained on a flat horizontal surface (surface plate) with the convex surface of the panel upward; the maximum vertical displacement shall be measured.

METHOD A. INDICATOR HEIGHT GAGE —

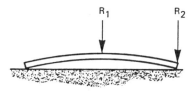

BOW

1. Place the printed board on a flat surface.
2. Take reading at edge of the printed board contacting flat surface R_2.
3. Take reading at maximum vertical displacement R_1.
4. Subtract R_2 from R_1.

NOTE: Care must be taken to make both readings on circuitry or base material.

5. Multiply length of printed board by allowable inch per inch bow. (Table I, II or III)
6. The difference of R_2 from R_1 should be equal to or less than the product in step 5.

METHOD B. FEELER GAGE —

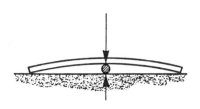

1. Place the printed board on a flat surface.
2. Multiply length or width of the printed board by inch inch bow. (Table I, II or III.)
3. Select feeler gage (gage blocks, or gage blanks), for the maximum allowance.
4. Attempt to slide feeler gage under printed board at point of greatest deviation.
5. If the feeler gage does not enter under the printed board the printed board is within tolerance. If gage enters under the printed board, the printed board is out of tolerance.

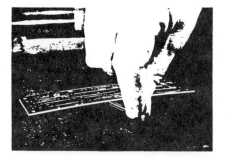

TWIST

Use either method described above, but measure diagonally (corner to corner) for inch per inch multiplier.

Fig. D-6 How to measure bow and twist on finished printed circuits. (Courtesy IPC)

IPC-TMM-650

2.1.1	Microsectioning
2.1.5	Surface Examination
2.2.7	Hole Size Measurement, Plated
2.2.9	Overhang & Undercut Measurement
2.2.10	Registration of Conductors
2.2.11	Registration, Terminal Pads (Layer to Layer)
2.2.13.1	Thickness, Plating in Holes, Micro-Ohm Method
2.3.4	Chemical Resistance, Legend Paints & Inks
2.3.11	Glass Fabric Examination
2.4.1	Adhesion, Plating
2.4.14	Solderability of Metallic Surfaces
2.4.22	Warp & Twist
2.5.4	Current Carrying Capacity, Multilayer Printed Boards
2.5.7	Dielectric Withstanding Voltage, Printed Board Material
2.5.16	Shorts, Internal on Multilayer Printed Boards
2.6.10	X-Ray (Radiology) Multilayer Printed Boards
5.8.1	Flexible Circuit Test Pattern
5.8.4	Rigid Multilayer Test Pattern

Fig. D-7   Tabulation of related test methods.

SECTION 14. FLAT CONDUCTOR, FLAT CABLE

Some areas of these types of circuitry are the same as those for printed circuits. These areas of commonality are listed, with a brief discussion.

SECTION 15. APPENDIX

This is a tabulation of test methods from IPC-TMM-650 (see Figure D-7). The appendix reproduces that actual test method.

Appendix E

MIL-P-13949F: Plastic Sheet, Laminated, Metal Clad (for Printed Wiring Boards)

This document is the industry standard for referencing materials for rigid printed circuit boards. It contains information on laminate and prepreg for all military recognized material types. The material designation is different from that of the National Electronics Manufacturing Association (NEMA) and must be understood; for example, instead of FR-4, the military designation is GF.

This specification has six sections plus an extensive appendix. The sections will not be fully discussed separately; but they are listed here: (1) Scope, (2) Applicable Documents, (3) Requirements, (4) Quality Assurance Provisions, (5) Packaging, (6) Notes, (7) Appendixes.

SECTION 1. SCOPE

Paragraph 1.1 states: "This specification covers the requirements for fully cured, metal-clad laminated, plastic sheets (glass and paper base) and semicured (B stage), resin-preimpregnated glass cloth (prepreg) to be used primarily for the fabrication of printed-wiring boards for electrical and electronic circuits (see 3.1 and 6.1). For the purposes of this specification, the term "laminate" will be used hereafter to denote metal-clad plastic sheets and the term 'prepreg' will be used to denote resin preimpregnated glass cloth (B stage). The term 'reinforced' will be used to denote a glass laminate and 'nonreinforced' will denote a laminate with no glass reinforcement."

The two materials, laminate and prepreg, each have their own type designation system. Also, in the back of this specification is a set of material specification sheets for each type of material. Designers often specify material on drawings as, for instance, MIL-P-13949F/4. This designation actually specifies GF material (NEMA designation: FR-4). This subject will be covered below.

1. Laminate, reinforced and nonreinforced, type designation.

GFN	0310	CH/D1	A	1	A
Base material	Nominal base thickness	Type and nominal weight of copper foil	Grade of pits and dents	Class of thickness tolerance	Class of bow and twist
(A)	(B)	(C)	(D)	(E)	(F)

a. Base materials are separated into 11 types. Unless otherwise designated, glass fiber is woven, compared to nonwoven, matte fibers.

   (1) PX—Paper base, epoxy resin, flame resistant.

   (2) GB—Glass base, epoxy resin, heat resistant, retains strength when hot.

   (3) GE—Glass base, epoxy resin, general purpose.

   (4) GF—Glass base, epoxy resin, flame retardant.

   (5) GH—Glass base, epoxy resin, heat resistant (retains strength when hot) and flame retardant.

   (6) GP—Glass base (nonwoven fiber), polytetrafluoroethylene resin, flame retardant.

   (7) GR—Glass base (nonwoven fiber), polytetrafluoroethylene resin, flame retardant, for microwave application.

   (8) GT—Glass base, polytetrafluoroethylene resin, flame retardant.

   (9) GX—Glass base, polytetrafluoroethylene resin, flame retardant, for microwave application.

  (10) GI—Glass base, polyimide resin, high temperature, heat resistant.

  (11) GY—Glass base, polytetrafluoroethylene resin, flame resistant, for microwave application.

The third letter (example: N in GFN) designation:

N—Base material without coloring agent or opacifier; frequently referred to as "natural."

P—Base material with coloring agent or opacifier. When P is specified on a drawing, as in "GFP," the planner or quality assurance person must look for a designated color, such as blue.

b. Nominal base thickness is designated by a four-digit number. This identifies the thickness to a ten-thousandth of an inch. Note that this excludes any metal cladding and refers strictly to the base (often referred to as "core") thickness.

  EXAMPLES: 0050 = .0050 inch = 5.0 mils

          : 0140 = .0140 inch = 14.1 mils

          : 0305 = .0305 inch = 30.5 mils

c. Type and nominal weight of copper foil five characters. There are two characters, a slash mark, and two more characters. Note that the slash mark is the third character. Example: CH/D1

(1) The letters C and D (first and fourth characters) indicate the type of copper foil cladding. Copper foil will be of the following types:

A — Rolled
B — Rolled (treated)
C — Drum side out, electrodeposited
D — Drum side out (double treated), electrodeposited
E — Matte side out, electrodeposited
F — Matte side out (double treated), electrodeposited
O — Unclad

(2) The H and 1 (second and fifth characters) indicate the nominal copper foil weight in ounces per square foot. If the foil cladding is 1 oz/sq ft or more, the actual ounce number shall be used in the designation. If the foil cladding is less than 1 oz/sq ft one of the letter indicators shown below shall be used.

Foil Thickness Designations

E—1/8 oz/sq ft	1 = 1 oz/sq ft
Q—1/4 oz/sq ft	2 = 2 oz/sq ft
T—3/8 oz/sq ft	
H—1/2 oz/sq ft	
O—unclad	

d. The grade of pits and dents is designated by letter A, B, or C. These grades differ by the point count allowed.

Grade A: The total point count in any 12 × 12-inch area shall be less than 30.
Grade B: The total point count in any 12 × 12-inch area shall be less than 30. There shall be no pits greater than .015 inch in the longest dimension. There shall be no more than three pits in a 12 × 12-inch area with the longest dimension over .005 inch.
Grade C: The total point count in any 12 × 12-inch area shall be less than 100.

Point count for pits and dents is as follows:

Longest Dimension (Inches)	Point Value
.005 to .010 inch incusive	1
.011 to .020 inch incusive	2
.021 to .030 inch incusive	4
.031 to .040 inch incusive	7
Over .040 inch	30

e. The class of thickness tolerance is designated by the number 1, 2, 3, or 4; these are the four classes. Thickness and tolerances do not apply to the outer 1.0 inch of a trimmed full-sized sheet. At least 90% of the sheet must meet the thickness class requirement, and no portion of the sheet shall vary more than 125% of the specified tolerance. Cut sheets of GP, GR, GT, GX, and GY shall meet the thickness tolerance requirement for the specified class in 100% of the area.

NOTE: The tolerance varies in each class with the nominal thickness of the laminate. Table E-1 lists the nominal thicknesses and tolerances.

f. The class of bow and twist is designated by letter A, B, C, or X. Classes A and B apply to full-size sheets of laminate. Class C applies only to laminate which has been cut to panel size. Class X indicates that bow and twist requirements are not applicable to single-sided laminates or to any other laminate with a base thickness of less than .020 inch. Table E-2 shows permissible bow and twist for full-size sheet classes A and B. Table E-3 shows the requirements for cut panels.

2. Prepregs and copper foils will not be discussed in depth here. However, Table E-4 is reproduced. This contains some information on glasses and prepregs. Table E-5 contains information on copper foil thickness and tolerance. Table E-5 is also reproduced here. Both the copper foil manufacturer and the laminate manufacturer must go through a certification process for their materials which is similar to the certification process called out under MIL-P-55110D for the printed circuit manufacturer. A discussion of this

Table E-1 Nominal Thickness and Tolerances for Laminates (Inch)

	Class		Class 2	Class 3[1]	Class 4
	Paper Base	*Glass*	*Glass*	*Glass*	*for Microwave*
Thickness	*Only*	*Reinforced*	*Reinforced*	*Reinforced*	*Application*
.0010 to .0045	—	±.0010	±.00075	±.0005[2]	—
.0046 to .0065	—	±.0015	±.0010	±.00075[2]	—
.0066 to .0120	—	±.0020	±.0015	±.0010[2]	—
.021 to .0200	—	±.0025	±.0020	±.0015[2]	—
.001 to .0299	—	±.0030	±.0025	±.0020[2]	—
.030 to .040[2]	±.0045	±.0065	±.0040	±.0030	±.002
.041 to .065[2]	⊥.0060	±.0075	±.0050	±.0030	±.002
.066 · to .100[2]	±.0075	±.0090	±.0070	±.0040	±.003
.101 to .140[2]	±.0090	±.0120	±.0090	±.0050	±.0035
.141 to .250[2]	±.0120	±.0220	±.0120	±.0060	±.0040

[1]These tighter tolerances are available only through product selection.
[2]Overall thickness including the copper foil (see 1.2.1.1.2).

Table E-2   Permissible Bow and Twist

	Total Variation, Maximum, Percent (on Basis of 36-Inch Dimension)[1]					
	Class A			*Class B*		
Thickness (Inch) (see Table E-1)	*All Types, All Weights Metal (One Side)*	*All Types, All Weights Metal (Two Sides)*		*All Types, All Weights Metal (One Side)*	*All Types, All Weights Metal (Two Sides)*	
		Glass	*Paper*		*Glass*	*Paper*
Over 0.020–	—	5	—	—	2	—
.030 or .031 –	12	5	6	10	2	5
.060 or .062 –	10	5	6	5	1	2.5
.090 or .093 –	8	3	3	5	1	2.5
.120 or .125 –	8	3	3	5	1	2.5
.240 or .250 –	5	1.5	1.5	5	1	1.5

[1]These values apply only to sheet sizes as manufactured and to cut pieces having either dimension not less than 18 inches.

[2]For nominal thicknesses not shown in this table (see 6.2), the bow or twist for the next lower thickness shown shall apply.

Table E-3   Bow and Twist or Cut-to-Size Panels

		Total Variation, Maximum, Percent			
		Laminate—Class C			
Thickness[2] *(inch) (See Table E-1)*	*Panel Size (Maximum Dimension, Inches)*	*All Weights of Foil, One Side*		*All Weights of Foil, Two Sides*	
		All Other Types	*Types GP, GR GT, GX, GY*	*All Other Types*	*Types GP, GR GT, GX, GY*
Over .020	8 or less	2.0	—	1.0	—
	8 to 12	2.0	—	1.5	—
	Greater than 12	2.5	—	1.5	—
.030 or .031	8 or less	1.5	3.0	.5	3.0
	8 to 12	1.5	3.0	1.0	3.0
	Greater than 12	2.0	3.0	1.0	3.0
.060 and over	12 or less	1.0	1.5	.5	1.5
	Greater than 12	1.5	1.5	.5	1.5

[1]Except when otherwise specified (see 3.1).

[2]For nominal thicknesses not shown in this table (see 6.2), the bow and twist for the next lower thickness shown shall apply.

process begins at paragraph 4.0. Only materials certified to MIL-P-13949F (or latest revision) should ever be used in printed circuits which must be certified to MIL-P-55110D.

Table E-5   Copper Thickness and Tolerance

Nominal Weight (oz/ft^2)	Tolerance by Weight (Percent)		Nominal[1] Thickness, Inches (Microns)	Tolerance[1] Inches (Microns)
	Class I	Class II		
$\frac{1}{8}$	±10	±5	.00020 (5.0)	—
$\frac{1}{4}$	±10	±5	.00036 (9.0)	—
$\frac{3}{8}$	±10	±5	.00052 (13.0)	—
$\frac{1}{2}$	±10	±5	.0007 (17.5)	±.0001 (2.5)
$\frac{3}{4}$	±10	±5	.0010 (25.0)	+.0002 (5.0)
1	±10	±5	.0014 (35.0)	±.0002 (5.0)
2	±10	±5	.0028 (70.0)	±.0003 (7.5)
3	±10	±5	.0042 (105.0)	±.0004 (10.0)
4	±10	±5	.0056 (140.0)	±.0006 (15.0)
5	±10	±5	.0070 (175.0)	±.0007 (17.5)
6	±10	±5	.0084 (210.0)	±.0008 (20.0)
7	±10	±5	.0098 (245.0)	±.0010 (25.0)
10	±10	±5	.0140 (350.0)	±.0014 (35.0)
14	±10	±5	.0196 (490.0)	±.0020 (50.0)

[1]Derives by weight test method 2.2.12 of IPC-CF-150.

3. Paragraph 4.8.2 begins a section on prepreg inspection that will be of use to quality assurance people. Areas of concern and testing include:
   a. Presence of dicyanodiamide crystals. See note at paragraph 6.11.
   b. Thread count, glass cloth thickness, and fabric weight.
   c. Resin gel time.
   d. Volatiles content.
   e. Resin content.
   f. Resin flow.
   g. Cured thickness.
   h. Electrical strength.
   i. Dielectric constant and dissipation factor.

4. Paragraph 4.8.3 begins a section on laminate inspection which will be of use to quality assurance people. Areas of concern and testing include:
   a. Pits and dents.
   b. Scratches.
   c. Solderability.
   d. Etch characteristics of the metal-clad surfaces.
   e. Plastic surface contamination (unclad side).
   f. Bow and twist.
   g. Thermal stress.
   h. Peel strength of foil
      • Before and after thermal stress.

- Before and after temperature cycling.
- At elevated temperatures.
- After exposure to process conditions.

i. Volume and surface resistivity.

j. Dimensional stability.

k. Water absorption.

l. Dielectric breakdown voltage.

m. Electrical strength.

n. Dielectric constant and dissipation factor.

15. PACKAGING

5.1 *Preservation.* Unless otherwise specified (see 6.2), clean and dry metal-clad laminates shall be interleaved with noncorrosive sheets to prevent abrasion. Unless otherwise specified (see 6.2), preimpregnated glass cloth shall be unit packaged and sealed in polyethylene bags in a manner that will afford adequate protection against corrosion, deterioration and physical damage during shipment and storage. The unit packaging shall be in a manner that will afford adequate protection against corrosion, deterioration, and physical damage during shipment from the supply source to the first receiving activity. This may conform to the contractor's commercial practice when such meets the requirements specified herein. The unit contractor shall be as specified in 5.2.

5.2 *Packing.* The metal-clad laminates and preimpregnated glass cloth shall be packed in shipping containers in a manner that will afford adequate protection against damage during direct shipment from the supply source to the first receiving activity. These packs shall conform to the applicable carrier rules and regulations and may be the contractor's commercial practice if these requirements are met.

5.3 *Marking.* In addition to any special marking required on the contract (see 6.2), each unit package and exterior container shall be marked with the following information (when applicable):

a. Specification number and type of material.

b. Manufacturer's material designation and lot number.

c. Quantity, unit of issue and roll or sheet dimensions.

d. Gross weight and cube.[1]

e. Date packed.[1]

f. Contract number.

g. Manufacturer's (contractor's) name and address.

h. Name and address of consignee.[1]

i. Cloth batch number and contractor's designation.[2]

j. Resin batch number and contractor's designation.[2]

k. Date of manufacture (impregnation) and manufacturer's recommended storage conditions (see 3.6.10).[2]

5.4 *General.* Exterior containers shall be of a minimum tare and cube consistent with the protection required and contain equal quantities of identical items to the greatest extent practicable.

---

[1] Required for shipping containers only.

[2] Required for resin preimpregnated glass cloth materials only.

Fig. E-1 Packaging requirements for materials.

o. Measurements of base materials GR and GX at X-band frequency.

p. Q (resonance) when applicable.

q. Flexural strength: ambient and elevated temperatures.

r. Arc resistance.

s. Flammability, when applicable.

SECTION 5. PACKAGING, COVERS PRESERVATION, PACKAGING, AND MARKING OF MATERIAL CERTIFIED TO MIL-P-13949F

This section is important and will be reproduced here. If a printed circuit manufacturer is buying materials which must be certified to MIL-P-13949F, all of these requirements must be met as the material comes into the shipping department (see Figure E-1).

SECTION 6. NOTES

This section contains useful notes, explanations, and definitions.

APPENDIXES

Appendix A: X-Band Effective Stripline Dielectric Constant and Dissipation Factor for Copper Clad Glass Woven Fabric GR and GX Laminate

Appendix B: Two Fluid, Three Terminal Method for the Measurement of Dielectric Properties at 1 MHz

Appendix C: Two Terminal, Contacting Electrodes Method for Measurement of Dielectric Properties

Appendix F ·

MIL-STD-2118: Design Requirements for Flex and Rigid-Flex Printed Wiring for Electronic Equipment

MIL-STD-2118 is one of the cornerstone documents in the world of flex and rigid-flex circuitry. If design, quality assurance, and planning personnel wish to become familiar with only one document to govern their flex programs, this is the one they should know. Many pitfalls in design can be avoided by following the design rules laid out here. No incoming/artwork inspection can be considered complete or accurate without a review of blueprints and artwork for conformance to the rules of this standard. The person who provides job quotations must perform a cursory engineering review to this standard in order to fully understand the consequences of what is being quoted.

This document is composed of six sections and an appendix: (1) Scope, (2) Referenced Documents, (3) Definitions, (4) General Requirements, (5) Detail Requirements, (6) Detail Part Mounting Requirements, and Appendix: Design Considerations.

SECTION 1. SCOPE

This section states the purpose of the standard and lists the classifications of flex circuitry. The purpose (from paragraph 1.1) is to accomplish the following:

1. To establish design requirements governing the following types of printed wiring:
   a. Flexible printed wiring, with or without shields or stiffeners.
   b. Rigid to flexible wiring (to be called "rigid-flex"), with or without plated-through holes.

2. To establish design considerations for mounting parts and assemblies thereon.

3. To establish the following requirements:
   a. For rigid-flex applications, conductor layers that are in the flexible portion are also to be considered conductor layers in the rigid multilayer board.
   b. All board types shall have the conductor patterns protected by an insulating layer, with two exceptions:
      (1) Land areas do not have to be covered by an insulating layer.
      (2) External conductors on the rigid portion of a type 4 rigid-flex circuit do not have to be covered by an insulating layer.
   c. When components are mounted on flex or rigid-flex circuits, the mounting area must be conformally covered in accordance with MIL-I-46058.

There are five types of circuitry, depending on the number of layers. Types 1, 2, and 3 correspond to types 1, 2, and 3 listed in MIL-STD-275 and MIL-P-55110D: single-sided, double-sided, and multilayer. Type 4 is rigid-flex multilayer with plated-through holes, and type 5 is rigid-flex multilayer without plated-through holes. The five types of circuitry are:

Type 1—Single-sided flex, with or without shields or stiffeners (one conductive layer).

Type 2—Double-sided flex, with or without shields or stiffeners (two conductive layers).

Type 3—Multilayer flex with plated-through holes, with or without shields or stiffeners (more than two conductive layers).

Type 4—Multilayer rigid and flex combinations with plated-through holes (more than two conductive layers).

Type 5—Bonded rigid or flex combinations without plated-through holes.

For each of the five types, there are two classifications:

Class A—Capable of withstanding flexing during installation.

Class B—Capable of withstanding continuous flexing for the number of cycles specified on the master drawing (generally not used for more than two conductive layers).

SECTION 2. REFERENCED DOCUMENTS

This section lists the titles and numbers of numerous federal, military, ANSI, and IPC specifications and standards. These are referenced throughout this standard. To fully understand all requirements of MIL-STD-2118, the reader must have a copy of these documents available for consultation.

SECTION 3. DEFINITIONS

This section does two things. It defines "splay" (which is the tendency of a rotating drill to drill holes which are off-center, out of round, or nonperpendicular) and it states that all other terms and definitions shall be in accordance with IPC-T-50.

SECTION 4. GENERAL REQUIREMENTS

1. Design Features. Paragraph 4.1 states that quality conformance test coupons shall be included on the production master, master drawing, and artwork. The coupons are shown in Figure F-1. The coupons are to be .250 to .500 inch from the board edges and must reflect all of the manufacturing processes which the boards themselves will undergo.

2. Master drawing. Paragraph 4.2 establishes the requirements to be met by the master drawing. These requirements include:
   a. Type, size, and shape of the flex or rigid-flex circuit.
   b. Size and location of all holes.
   c. Whether or not etchback is required.
   d. Location of traceability markings.
   e. Dielectric separation between layers.
   f. Number and location of quality conformance test coupons.
   g. Shape and arrangement of conductor and nonconductor patterns or elements.
   h. Separate views of each conductor layer for flex and rigid-flex layers.
   i. All pattern features not controlled by the hole size and locations shall be adequately dimensioned, either by specific dimensions or by notes on the blueprint.
   j. Step or repeat of circuit patterns or quality conformance test coupons (that is, generation of working film) shall meet the requirements for production master accuracy (paragraph 4.3).
   k. Definitions of all terms used on the drawing shall be in accordance with IPC-T-50.
   l. Plating and coating thickness shall be specified on the master drawing.
   m. The master drawing shall list all artwork considerations that were used in the design of the flex or rigid-flex printed circuit. This means that processing tolerances (minimum and maximum values) for trace width, conductor spacing, annular ring, etc. shall be noted somewhere on the

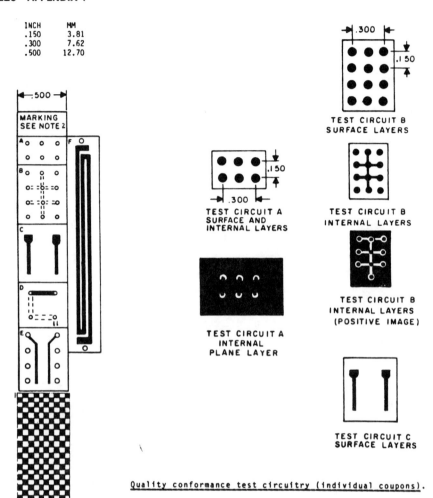

INCH	MM
.150	3.81
.300	7.62
.500	12.70

TEST CIRCUIT B
SURFACE LAYERS

TEST CIRCUIT A
SURFACE AND
INTERNAL LAYERS

TEST CIRCUIT B
INTERNAL LAYERS

TEST CIRCUIT A
INTERNAL
PLANE LAYER

TEST CIRCUIT B
INTERNAL LAYERS
(POSITIVE IMAGE)

TEST CIRCUIT C
SURFACE LAYERS

MARKING
SEE NOTE 2

Quality conformance test circuitry (individual coupons).

Quality conformance test circuitry layout.

Fig. F-1   Quality conformance test circuitry.

master drawing. This standard recognizes that the end product will not match the artwork exactly and that the allowable variations must be spelled out. All feature detail requirements (to be discussed in Section 5) shall be spelled out.

3. Number of Sheets for the Master Drawing
   a. If practical, the master drawing shall be kept to one sheet.
   b. Requirements for multisheet master drawings

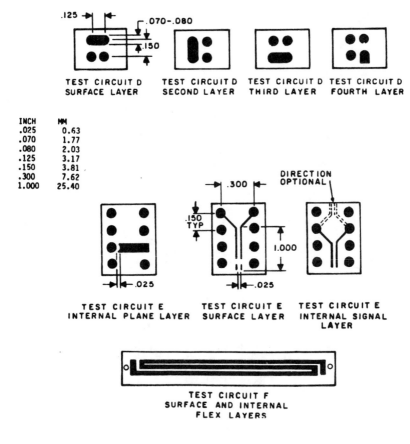

TEST CIRCUIT D
SURFACE LAYER

TEST CIRCUIT D
SECOND LAYER

TEST CIRCUIT D
THIRD LAYER

TEST CIRCUIT D
FOURTH LAYER

INCH	MM
.025	0.63
.070	1.77
.080	2.03
.125	3.17
.150	3.81
.300	7.62
1.000	25.40

TEST CIRCUIT E
INTERNAL PLANE LAYER

TEST CIRCUIT E
SURFACE LAYER

TEST CIRCUIT E
INTERNAL SIGNAL
LAYER

TEST CIRCUIT F
SURFACE AND INTERNAL
FLEX LAYERS

Quality conformance test circuitry (individual coupons).

Fig. F-1   *(Continued)*

The first sheets shall establish the size and shape of the printed circuit, stiffeners, hole diameters, tolerances and locations, and all notes. Pattern features not controlled by hole locations shall be dimensioned specifically, or by notes, on these first sheets.

c. Subsequent sheets shall establish the shape and arrangement of conductor and nonconductor layers.

d. Conductor layers are to be numbered sequentially from the component side. The first conductor layer is to be layer 1.

e. Locations of manually applied markings (scribed or hand stamped) shall be described in notes.

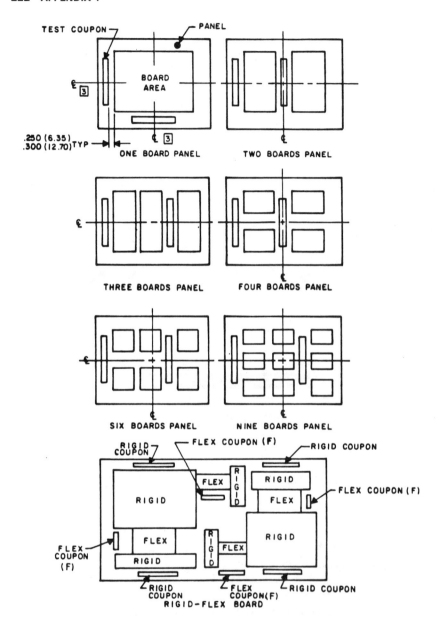

Typical location of test coupon based on number of boards fabricated per panel.

Fig. F-1   (*Continued*)

NOTES:

1. Dimensions are in inches.
2. Test coupons are to be identified with the following:
   a. FSCM.
   b. Part number and revision letter.
   c. Board traceability or lot number.
3. All lines shall be .020 (0.51 mm) ±.003 (0.08 mm), unless otherwise specified.
4. Unless otherwise specified, the tolerances shall meet the requirements of this standard.
5. The minimum land dimensions shall be .070 (1.78 mm) ±.005 (0.13 mm) and represent the land shape used on the associated board. Holes in lands shall be the diameter of the smallest component hole in the associated board.
6. All first layers and internal layers shall be as specified on the master drawing. Copper plane areas shall be used on all coupons on appropriate plane layers, except for the D and E segments. When shields are used (see 5.11) appropriate layers shall be added to the coupons.
7. The lengths of test circuits D and E are dependent upon the number of layers in the panel. For test circuits D, a pair of holes and a conductor between same shall be provided for each layer. Electrical connection shall be in series, stepwise, through each conductor layer of the board. For test circuit E, a pair of holes and conductors shall be provided for the first layer and each internal layer.
8. Coupon F shall be positioned in the flexible area on the associated board.
9. The quality conformance test circuitry may be segmented; however, test circuitry A and B shall be joined together. Test circuit C, D, E, and F may be arranged to optimize board layout. All test coupons illustrated shall appear on each panel. The number of layers shall be identical to the number of layers in the boards derived from the panel.
10. Letters on coupons are for identification purposes only and shall be etched or applied by the use of a permanent ink which will withstand board processing. Location of letters on applicable coupons is optional.
11. Number of layers shown in these test coupons are for illustration purposes only. Conductor layer number 1 shall be the first layer on the component side, and all other conductor layers shall be counted consecutively downward through the laminated board to the bottom conductor layer which is the solder side. Surface layers shall consist of the outer layers of a printed-wiring board, the first layer (component side) and the last layer (solder side).

Fig. F-1 *(Continued)*

f. Photographically applied markings (legend and solder mask) shall have locations and methods specified on a separate sheet, preferably the last sheets in the set.

4. Other Requirements for the Master Drawing
   a. Location dimensioning. Holes, test points, lands, and overall dimensions shall be specified by the use of a grid pattern. The grid shall be .100, .050, or inch .025 or another multiple of .005 inch. There is other information in paragraph 4.2.3 which is of interest to a designer.
   b. Hole patterns. Each distinctive set of holes shall be dimensioned from a primary or secondary grid system. Types of holes include plated-through holes, tooling holes, mounting holes, windows, access holes, etc. Each

Table F-1 Board Design Guidelines

	Preferred	Standard	Reduced Producibility
Number of conductive layers (types 3 and 4)	6	12	20
Composite thickness tolerance (types 1, 2, 3, and 4)	+, −20% of nominal	+, −15% of nominal	+, −10% of nominal or +, −0.010 inch, whichever is greater
Thickness of dielectric rigid materials—type 4 (min.)	.008 (.20)	.006 (.15)	.0035 (.089)
Flex-polymide—types 1, 2, 3, and 4 (min)	.002 (.05)	.002 (.05)	.0015 (.038)
Stiffeners	.031 (.79)	.062 (.6)	.090 to .125 (2.3 to 3.2)
Minimum conductor width (or Figure 4 value, whichever is greater)			
Internal	.15 (0.38)	.008 (0.20)	.004 (.10)
External	.20 (0.51)	.015 (0.38)	.008 (.20)
Conductor width tolerance			
Unplated 2 oz/ft^2	+.004 (.10) −.006 (.15)	+.002 (.05) −.005 (.13)	+.001 (0.025) −.003 (0.08)
Unplated 1 oz/ft^2	+.002 (.05) −.003 (.08)	+.001 (.025) −.002 (.05)	+.001 (.025) −.001 (.025)
Unplated 1/2 oz/ft^2	—	—	+.0005/−.001 (.013/.03)
Metallic etch resist in external layers			
Over 2 oz/ft^2 Cu	+.008 (.20) −.006 (.15)	+.004 (.10) −.004 (.10)	+.002 (.05) −.002 (.05)
Minimum conductor spacing (or Table 1, whichever is greater)	.015 (.38)	.008 (.20)	.004 (.10)
Annular ring plated-through hole (min.) (See 5.2.2)			
Internal	.008 (.20)	.005 (.13)	.002 (.05)
External	.010 (.25)	.008 (.20)	.005 (.13)
Annular ring unsupported hole, rigid and flex (min)	.020 (.51)	.015 (.38)	.010 (.25)
Feature location tolerance (rtp) (types 1, 2, 3, and 4) (master pattern and registration)			
Longest dimension (12 inches or less)			
Feature pattern (rtp) area of types 3 and 4	.006 (.15)	.005 (.13)	.004 (.10)
Feature pattern (rtp) area of types 1 and 2	.015 (.38)	.010 (.25)	.008 (.20)

Table F-1   (Continued)

	Preferred	Standard	Reduced Producibility
Longest dimension (over 12 inches) (rtp)			
Feature pattern (rtp) area of types 3 and 4	.008 (.20)	.007 (.18)	.006 (.15)
Feature pattern (rtp) area of types 1 and 2	.025 (.64)	.020 (.51)	.015 (.38)
Predicted on total board configuration (see 4.2.6)			
Master pattern accuracy (rtp)			
Longest board dimension 12 inches or less	.004 (.10)	.003 (.08)	.002 (.05)
Longest board dimension over 12 inches	.005 (.13)	.004 (.10)	.003 (.08)
Feature size tolerance	+, −.003 (.08)	+, −.002 (.05)	+, −.001 (.025)
Composite thickness to hole diameter	3 : 1	4 : 1	5 : 1
Rigid (types 3 and 4) max.			
Hole location tolerance (rtp)			
Component mounting area— dimension of 12 inches or less	.007 (.18)	.005 (.13)	.003 (.08)
Component mounting area— dimension of over 12 inches	.010 (.25)	.008 (.20)	.005 (.13)
Unplated hole diameter tolerance (unilateral)			
Up to .032 (.81)	.004 (.10)	.003 (.08)	.002 (.05)
.033 .063	.006 (.15)	.004 (.10)	.002 (.05)
.064 .188	.008 (.20)	.006 (.15)	.004 (.10)
Plated hole diameter tolerance (unilateral); for minimum hole diameter maximum board thickness ratios greater than 1.4, add .004 (.10) inch			
.015 (.38) .030 (.76)	.008 (.20)	.005 (.13)	.003 (.08)
.031 (.79) .061 (1.56)	.010 (.25)	.006 (.15)	.004 (.10)
.062 (1.59) .186 (4.75)	.012 (.31)	.008 (.20)	.006 (.15)
Conductor, including terminal pads to edge of composite (min)			

Table F-1   (*Continued*)

	Preferred	Standard	Reduced Producibility
Conductor, including terminal pads to edge of composite (min) (*Continued*)			
Types 3 and 4	.100 (2.54)	.075 (1.90)	.050 (1.27)
Types 1 and 2	.100 (2.54)	.050 (1.27)	.025 (.63)
Bending of flex (internal radius) thickness			
Class A	6× material	1× material	
.003 to .010	thickness	thickness	
(.08 to .25)			
Over .010 (.25)	12× material	2× material	1× material
	thickness	thickness	thickness
Class B			
.020 max (.51)	1 inch	1/2 inch	1/4 inch
Continuous flex cycles (based 1 oz/ft² copper, .002 inch dielectric) thickness			
.003 to .008			
(.08 to .20)	1000	100,000	500,000
Over .008 to .014			
(0.20 to 0.36)	500	50,000	250,000
Over .014 (.36)	250	25,000	100,000

NOTE: Unless otherwise specified, all dimensions and tolerances are in inches. Data in parentheses ( ) are expressed in millimeters.

NOTE: This figure is not shown. The reader should reference MIL-STD-2118.

set may require separate dimensional tolerance considerations. Table F-1, which summarizes design guidelines, is reproduced here. These guidelines are based, like other military specifications, on producibility. Preferred, Standard, and Reduced Producibility are the three classes.

c. Data. There shall be a minimum of two mutually perpendicular datum lines. Critical features with critical location requirements may be referenced by a secondary datum.

d. Government-furnished master drawings must be in accordance with this standard. Any deviations must be recorded on the government-approved master drawing.

5. Assembly Drawings. Requirements for assembly drawings are discussed beginning with paragraph 4.4. It is common for manufacturers of flex and rigid-flex to be required to perform some assembly. For this reason, it is a good idea to know the basic drawing requirements. Minimum requirements include:

a. Lead-forming requirements.

b. Cleanliness requirements per MIL-P-28809.

c. Types of materials (conformal contain, masking, and potting).

d. Location and identification of all components.

e. Component orientation and polarity.

f. Applicable ordering data from MIL-P-28809.

g. Structural details, when required, for support and rigidity.

h. Electrical circuitry test requirements.

i. Marking requirements.

j. All appropriate assembly requirements shall be listed and defined on the assembly drawing, including allowances and necessary manufacturing data.

SECTION 5. DETAIL REQUIREMENTS

This section covers requirements for conductors, spacing, test points, cover sheets, bends, holes, plated metals, and solder mask. It is important that planners, designers, and quality assurance people have an understanding of this material.

1. Conductor Pattern

a. Conductor thickness and width shall be determined in accordance with Figure F-2. Widths and spacings shall be maximized for manufacturability and operating durability, consistent with good design pratice. The minimum allowable conductor width is .004 inch. Working film (production master) shall be modified to compensate for processing tolerances. Conductor width tolerances are listed in Table F-1.

b. Conductors with exterior angles less than 90 degrees shall be rounded.

c. Conductor lengths shall be held to a minimum, with the following considerations:

(1) Conductors shall run parallel with the X or Y axis directions or at a 45 degree angle to these directions.

(2) Parallel conductors on opposite sides of a dielectric that will be flexed shall be offset from each other.

(3) Flexing, or bending, shall be allowed only in areas where the conductors are straight and the bend radius is at 90 degrees to the conductor paths (see Figure F-3).

d. Conductor spacing shall be as large as possible. Minimum conductor spacing shall be in accordance with Table F-2.

e. Jumper wires are permitted only with prior approval of the procuring agency. Paragraph 5.1.5 lists other requirements for jumper wires.

f. Minimum edge spacing from any conductive feature to the board edges shall be per Table F-2—only if the edges are protected from physical harm in the installed assembly. Otherwise, the minimum conductor-to-

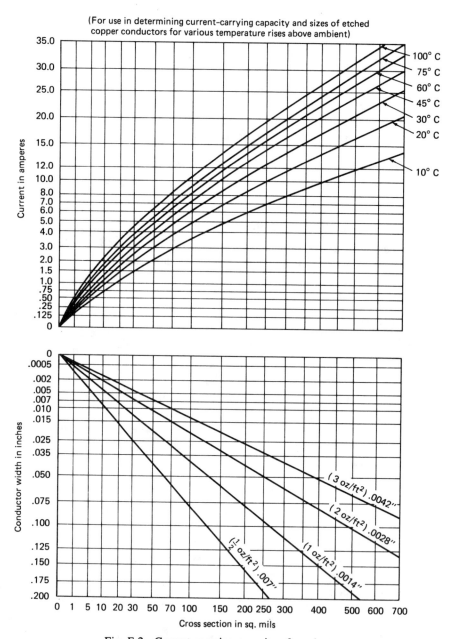

Fig. F-2 Current-carrying capacity of conductors.

NOTES:

1. The design chart has been prepared as an aid in estimating temperature rises (above ambient) vs. current for various cross-sectional areas of etched copper conductors. It is assumed that normal design conditions prevail where the conductor surface area is relatively small compared to the adjacent free panel area. The curves as presented include a nominal 10% derating (on a current basis) to allow for normal variations in etching techniques, copper thickness, conductor width estimates, and cross-sectional area.
2. Additional derating of 15% (current-wise) is suggested under the following conditions:
    (a) For panel thickness of 1/32 inch or less.
    (b) For conductor thickness of 0.0042 inch (3 oz/ft^2) or thicker.
3. For general use, the permissible temperature rise is defined as the difference between the maximum safe operating temperature of the laminate and the maximum ambient temperature in the location where the panel will be used.
4. For single conductor applications the chart may be used directly for determining conductor widths, conductor thickness, cross-sectional area, and current-carrying capacity for various temperature rises.
5. For groups of similar parallel conductors, if closely spaced, the temperature rise may be found by using an equivalent cross section and an equivalent current. The equivalent cross section is equal to the sum of the cross sections of the parallel conductors, and the equivalent current is the sum of the currents in the conductors.
6. The effect of heating due to attachment of power-dissipating parts is not included.
7. The conductor thicknesses in the design chart do not include conductor overplating with metals other than copper.

Fig. F-2   (*Continued*)

board edge shall be .100 inch. Edge spacing does not apply to shield/ ground planes or heat sinks.
g. Large external conductive areas, exceeding 1.0 inch in diameter, shall be broken up by etched-out areas. Large conductive areas should be on the component side of type 3 and type 4 circuits.
h. Large internal conductive areas, exceeding 1.0 inch in diameter, shall be broken up by etched-out areas and located near the center of the board. If more than one such layer is used, these layers must be located to provide for balanced construction.
i. Interfacial connections on type 2 boards can be made by plated-through holes or clinched wires. Clinched wires being used to make interfacial connections are not considered part of the assembly. Interfacial connections on type 3 and 4 boards shall be made only with plated-through holes.
j. Solder plugs in holes which do contain leads are permissible after wave soldering. See paragraph 5.1.8.1 for other rules about solder plugs.

2. Lands shall be provided at each location where a part lead occurs or at any other location where an electrical connection is made, including a test point.

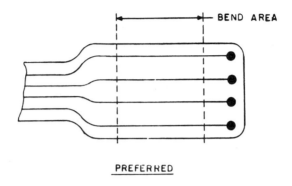

PREFERRED

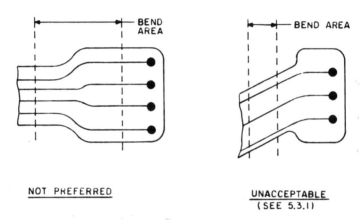

NOT PREFERRED

UNACCEPTABLE
(SEE 5.3.1)

Fig. F-3   Bending rules for conductors.

Table F-2   Conductor Spacing

Voltage Between Conductors DC or AC Peak (Volts)	Minimum Spacing (Inches)	
	Surface	Encapsulated[2]
0–100	.005 (.13 mm)	.004 (.10 mm)
101–300	.015 (.38 mm)	.008 (.20 mm)
301–500	.030 (.76 mm)	.010 (.25 mm)
Greater than 500[1]	.00012 (.0030 mm) per volt	.0001 (.003 mm) per volt

[1]For reference only, voltage greater than 500 volts should be evaluated for the specific design application.

[2]"Encapsulated" means the internal layers bonded together or the external layers with cover coat or potting, as opposed to conformal coating or solder mask.

Lands for surface-terminated flat packs shall be rectangular (preferably) and shall have the following dimensions:
a. Minimum width: equal to or greater than the maximum lead width.
b. Minimum length: equal to or greater than twice the minimum width.

3. Etchback. When required, etchback shall be specified on the master drawing. It shall be at least .0001 inch minimum and .003 inch maximum.

4. Negative etchback at the internal layers shall be allowed to a maximum of .003 inch.

5. Land areas shall be as large as possible. Minimum land areas shall be calculated as follows:

$$\text{Minimum land area} = a + 2b + 2c \text{ (when required)} + d$$

a = Maximum diameter of the drilled hole for internal lands or maximum diameter of the finished hole for external lands.
b = Minimum annular ring requirement.
c = Maximum allowance for etchback, when required.
d = Standard fabrication allowance, determined by statistical survey. See Table F-3 for allowable standard fabrication tolerances which can be used.

6. Annular ring considerations. The minimum annular ring for flexible circuits is determined in the same manner as for rigid printed circuits.
a. External measurements—from the inside edge of the hole to the outside edge of the land at the narrowest location.
b. Internal measurements—from the inside edge of the drilled hole wall to the outside edge of the land at the narrowest location.
Minimum requirements for the annular ring are as follows:

Table F-3 Standard Fabrication Allowances

Greatest Board Dimension	Allowances (inches)		
	Preferred	Standard	Reduced Producibility
Up to 12 inches	.028 (.71 mm)	.020 (.51 mm)	.012 (.30 mm)
12 to 18 inches	.034 (.86 mm)	.024 (.61 mm)	.016 (.41 mm)
More than 18 inches	Drawing tolerances must reflect bend and fold allowances between component mounting rigid areas.		

    a. External
      (1) Minimum annular ring for plated holes on types 2, 3, and 4 is .005 inch.
      (2) Minimum annular ring for nonplated holes is .015 inch.
    b. Internal
      (1) Minimum annular ring at functional lands is .002 inch.

7. Lands in large conductive planes shall be relieved according to the examples shown in Figure F-4.

8. Bends
    a. Bends should be kept to a minimum.
    b. Conductors must be perpendicular to bend lines.
    c. Component holes, plated-through holes, and surface mounting lands must be at least .100 inch from a bend.
    d. Class B, continuous flexing applications must not have plating in the bend area.

9. The bend radius should be as large as possible. Suggested minimum allowed radii are:
    a. Type 1 and 2 boards—minimum radii should be at least 6 times the maximum overall thickness.
    b. Type 3, 4, and 5 boards—minimum radii should be at least 12 times the maximum overall thickness.

10. Strain reliefs should be used, when applicable, to relieve pressure on solder joints.

11. Prebending should be avoided, per paragraph 5.3.3.

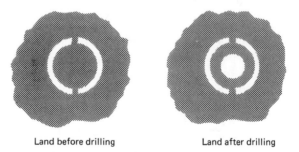

Land before drilling   Land after drilling

Fig. F-4   Ground plane lands.

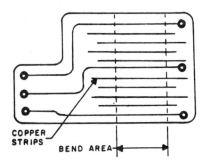

Fig. F-5   Method of increasing bend strength.

12. Bend strengtheners are permissible to increase the strength of circuits when bends occur in an area with few conductors. Bend strengtheners are strips of copper added in parallel with conductors (see Figure F-5).

13. Periphery of circuits (see Figure F-6). Good rules to follow are the following:
    a. Keep the shape as simple as possible.
    b. Avoid sharp corners.
    c. Inside corners and slots should have tear stop holes or copper dams added.
    d. Outside corners may be chamfered or formed with a radius (minimum radius, .015 inch).

14. Cover sheets and access holes. Access holes are holes in the cover sheets which allow access to the circuitry or other holes. Good design rules are as follows:
    a. Access holes should have a diameter at least .030 inch greater than the diameter of the component hole in the conductive layer.

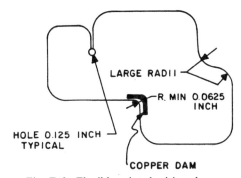

Fig. F-6   Flexible printed-wiring shape.

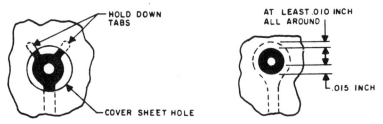

Fig. F-7 Coverlay sheet must capture pads.

b. Pads must be held down by the cover sheet. If the cover sheet does not overlap at least .010 inch on the pads of unsupported holes, anchoring spurs must be added to those pads (see Figure F-7).

c. When using reduced producibility per Table F-3 (standard fabrication allowances), all lands shall have anchoring spurs added.

d. Forms of access holes for tightly spaced holes (see Figure F-8). When individual access holes are not practical, the access holes may take the form shown in Figure F-8. However, unsupported holes must have anchoring spurs added.

15. Holes of various types will now be discussed.

a. For unsupported holes, the maximum diameter should not be .020 inch greater than the diameter of the lead to be inserted. When a rectangular pin is to be inserted, the hole diameter shall not exceed the nominal diagonal of the pin by more than .028 inch.

b. Holes which will have eyelets inserted shall have diameters no more than .010 inch greater than the outside diameter of the eyelet. The inside diameter of the eyelet shall not be more than .028 inch greater than the lead which will be inserted.

c. Indexing holes must be dimensioned on the master drawing.

d. Plated-through holes

(1) Plated-through holes shall not be used for mounting eyelets or standoffs.

(2) Plated-through holes which will have leads terminating in them shall be solder coated or plated. The minimum solder thickness shall be .0001 inch.

16. Flexible metal-clad materials shall be per IPC-FC-241, with a minimum base thickness of .001 inch. Other metal-clad materials shall be GF or GI per MIL-P-13949, with a minimum base thickness of .002 inch (see Table F-4).

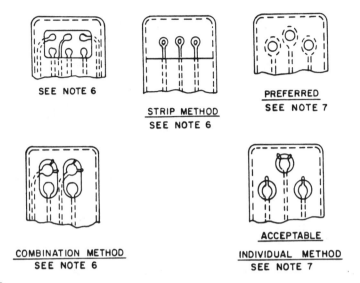

SEE NOTE 6

STRIP METHOD
SEE NOTE 6

PREFERRED
SEE NOTE 7

COMBINATION METHOD
SEE NOTE 6

ACCEPTABLE

INDIVIDUAL METHOD
SEE NOTE 7

NOTES:

1. The combination and individual methods are the most costly. The strip method creates a weak spot where the copper and the base material may crack.
2. The individual method shall be used for flexible printed-wiring with low density lands (>.15 centers).
3. Strip or combination methods shall be used for flexible printed-wiring with high density lands (<.15 centers).
4. Strip method (baring conductors) shall always be encapsulated at assembly and be provided with strain relief.
5. Combination method (baring conductors) shall always be conformal coated or encapsulated at assembly.
6. For use with plated through holes only.
7. For use with unsupported and plated through holes.

Fig. F-8   Forms of access holes.

Table F-4   Flexible Metal-Clad Dielectric

Specification Sheet	Material Identification
IPC-FC-241/1	Copper clad, polyimide with acrylic adhesive
IPC-FC-241/2	Copper clad, polyimide with epoxy adhesive
IPC-FC-241/3	Copper clad, fluorinated poly (ethylene-pro-plylene) (FEP) with acrylic adhesive
IPC-FC-241/4	Copper clad, fluorinated poly (ethylene-pro-plylene) (FEP) with epoxy adhesive

Table F-5   Cover Layer

Specification Sheet	Material Identification
IPC-FC-232/1	Polyimide base dielectric with acrylic adhesive
IPC-FC-232/2	Polyimide base dielectric with epoxy adhesive
IPC-FC-232/3	FEP base dielectric (fluorinated poly [ethylene proplyene]) with acrylic adhesive
IPC-FC-232/4	FEP base dielectric (fluorinated poly [ethylene proplyene]) with epoxy adhesive

17. Coverlayer material shall be per IPC-FC-232 and Table F-5. The minimum thickness of the base plus adhesive shall be .001 inch.

18. Adhesives (see Table F-6)
   a. Prepreg shall be GE, GF, or GI per MIL-P-13949. Areas requiring prepreg or adhesive shall be identified in the master drawing.
   b. Flexible adhesive bonding films shall be in accordance with IPC-FC-233. Locations requiring adhesives are to be identified on the master drawing.
   c. Adhesives and prepregs can be used interchangeably in the rigid portion of type 4 boards, provided dielectric requirements are met.

19. Fillets of adhesive may be required as strain reliefs at junctures of stiffeners or rigid-flex junctures. Requirements of the fillet shall be on the master drawing (see Figure F-9).

20. Stiffeners and heat sink materials. Thickness and adhesive shall be specified on the master drawing when required. Registration requirements of stiffener access holes to the printed circuit shall also be specified on the master drawing (see Figure F-10). Stiffeners may be internal or external. The edge of the stiffener shall be chamfered or rounded to prevent damage to conductors where it meets the flexible portion of the circuit (see Figure F-11).

21. Copper circuitry layers shall be in accordance with IPC-FC-150. A minimum of $\frac{1}{2}$ oz/sq ft is required for external layers.

Table F-6   Adhesives

Specification Sheet	Material Identification
IPC-FC-233/1	Acrylic adhesive
IPC-FC-233/2	Epoxy adhesive

[FLEXIBIE DIELECTRIC]

[COPPER]

[ADHESIVE]

[SOLDER OR (FUSED) TIN-LEAD PLATE]

[RIGID DIELECTRIC (USUALLY CONTAINS GLASS)]

MATERIAL LEGEND

WITHOUT STIFFENER

WITH STIFFENER

STRAIN RELIEF
FILLET-WHEN USED

Single-sided flexible printed wiring (type 1).

WITHOUT STIFFENER

Fig. F-9 Typical constructions of flexible and rigid-flex printed circuits.

<u>WITH STIFFENER</u>

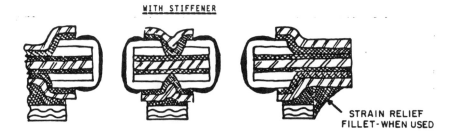

STRAIN RELIEF
FILLET-WHEN USED

<u>Double-sided flexible printed-wiring (type 2).</u>

<u>FLEXIBLE SECTION WITHOUT STIFFENER</u>

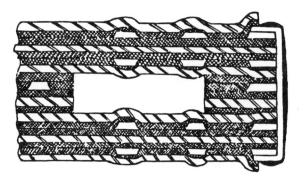

Five layer depicted - 2 Double sided with coverlayers
plus 1 Single layer (encapsulated)

<u>FLEXIBLE SECTION WITH STIFFENER</u>

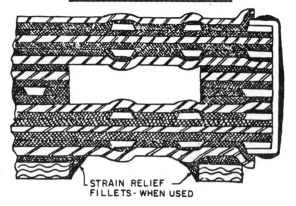

STRAIN RELIEF
FILLETS- WHEN USED

<u>Multilayer flexible printed-wiring (type 3).</u>

Fig. F-9   (*Continued*)

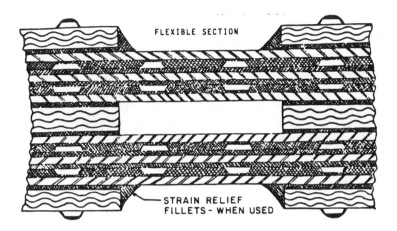

FLEXIBLE SECTION

STRAIN RELIEF
FILLETS - WHEN USED

PLATED THROUGH HOLE SECTION

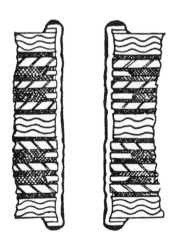

NOTE:

Varying degrees of preferential etchback
amongst the various materials, as depicted
is to be expected in finished product.

Multilayer composite combination rigid-flex printed-wiring (type 4).

Fig. F-9 (*Continued*)

    a. Flexible layers
- Type W, class 7

    b. Rigid layers
- Type E, class 1, 2, 3, or 4

or

- Type W, class 7

    c. Class B flexible, and flexible portion of rigid-flex boards
- Type w, class 7

22. Plating
    a. Electroless copper—for all plated-through holes.
    b. Electrolytic copper—per MIL-C-14550, with a minimum thickness of .001 inch.
    c. Gold—per MIL-G-45204, with a minimum thickness of .000050 inch over low-stress nickel.
    d. Tin/lead—per MIL-P-81728; it shall have a minimum fused thickness of .0003 inch at the crest of the conductor.

23. Solder mask shall be limited to the rigid portions of type 4 boards and shall meet IPC-SM-840, class 3, requirements when specified.
    a. If the mask is to cover tin-lead, metal areas larger than .050 in^2 should be relieved with at least .010 in^2. This promotes solder mask adhesion.
    b. Metal areas which are to be uncovered shall have solder mask overlapping edges by .010 to .250 inch.

24. Flex and rigid-flex thickness tolerances
    a. Keep tolerances as liberal as possible. If a thickness tolerance is required, limit it to the portion where thickness control is critical.
    b. When parts are being mounted, measure across metal extremities.
    c. When base material thickness is critical, measure across the dielectric only.

25. Minimum dielectric thicknesses
    a. In type 4 circuits constructed with MIL-P-13949 materials, minimum dielectric spacing between conductive layers shall be .0035 inch. Dielectric includes prepreg and copper-clad material. There shall be at least two sheets of glass material between the conductive layers.
    b. Circuits constructed using flexible materials and adhesives shall have at least .0015 inch of dielectric spacing. Spacing includes the flexible base plus adhesive.

26. Bow and twist shall be 1.5% of the rigid section of type 4 circuits only.

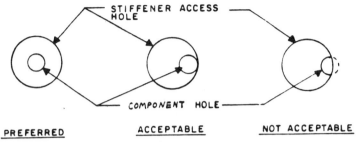

Fig. F-10   Stiffener registration.

Fig. F-11   Radiused edge of stiffener.

27. Shielding materials other than copper foil shall be approved by the procuring agency and included on the quality conformance test coupons and master drawings. Examples of shielding materials are silver epoxy and vapor-
deposited metals.

SECTION 6. DETAIL PART MOUNTING REQUIREMENTS

This section begins on page 14 of MIL-STD-2118 and will not be covered here.

APPENDIX

The Appendix contains a few paragraphs on design considerations and Table
F-1.

Index